ROUTLEDGE LIBRARY EDITIONS: URBAN PLANNING

Volume 1

URBAN LAND AND PROPERTY MARKETS IN FRANCE

URBAN LAND AND PROPERTY MARKETS IN FRANCE

RODRIGO ACOSTA AND VINCENT RENARD

Routledge
Taylor & Francis Group

LONDON AND NEW YORK

First published in 1993 by UCL Press

This edition first published in 2018
by Routledge
2 Park Square, Milton Park, Abingdon, Oxon OX14 4RN

and by Routledge
711 Third Avenue, New York, NY 10017

Routledge is an imprint of the Taylor & Francis Group, an informa business

British Library Cataloguing in Publication Data
A catalogue record for this book is available from the British Library

ISBN: 978-1-138-49611-8 (Set)
ISBN: 978-1-351-02214-9 (Set) (ebk)
ISBN: 978-1-138-49536-4 (Volume 1) (hbk)
ISBN: 978-1-351-02414-3 (Volume 1) (ebk)

Publisher's Note
The publisher has gone to great lengths to ensure the quality of this reprint but points out that some imperfections in the original copies may be apparent.

Disclaimer
The publisher has made every effort to trace copyright holders and would welcome correspondence from those they have been unable to trace.

Urban land and property markets in France

Rodrigo Acosta
Association pour le Développement des Études Foncières

Vincent Renard
*Association pour le Développement des Études Foncières
and Centre National de la Recherche Scientifique*

UCL
PRESS

First published in 1993 by UCL Press

UCL Press Limited
University College London
Gower Street
London WC1E 6BT

The name of University College London (UCL) is a registered
trade mark used by UCL Press with the consent of the owner.

ISBN:
1-85728-050-4 HB

British Library Cataloguing-in-Publication Data
A catalogue record for this book
is available from the British Library.

Typeset in Times Roman.
Printed and bound by
Biddles Ltd, King's Lynn and Guildford, England.

CONTENTS

FOREWORD TO SERIES

The idea of publishing this series of books on the different national urban land and property markets of Europe was inspired by a five-country research project on the functioning and framework of urban land and property markets. This project, known as the EuProMa Project, was commissioned by the German Federal Ministry for Regional Planning, Building and Town Planning (Bundesministerium für Raumordnung, Bauwesen und Städtebau, or BMBau), and was undertaken at the Faculty of Spatial Planning (Fakultät Raumplanung) of the University of Dortmund, Germany, under the direction of Hartmut Dieterich who holds the Chair in Vermessungswesen und Bodenordnung.

There is a growing interest in the land and property markets throughout Europe. The Single European Market (SEM) is to become a reality as early as January 1993, as is the European Economic Area extending SEM benefits to the EFTA countries. Land use and urban development will be influenced by the SEM in many ways, and competition between regions and cities will grow. Furthermore, ratification of the Treaty of European Union or Maastricht Treaty, (which explicitly refers to "town and country planning" and "land-use" in Article 130s(2)) is expected to lead to more economic and environmental regulations which will be important factors in the operation of the land market.

The central objective of the EuProMa project was to prepare detailed accounts of the actual operations of the urban land and property markets in five major EC economies. This was necessarily supported by accounts of the planning, taxation and legal framework within which the markets operate.

France, Italy and the United Kingdom were selected because of their importance within the EC, as well as their importance for Germany as trading partners. The Netherlands was of interest because Dutch developers are increasingly active in the German property market, a consideration which is important for Britain also, since many British real estate firms and developers have already set up shop in Germany.

The BMBau commissioned the project because it wanted to know how the land and property market functions elsewhere, why it is as it is, whether the outcome of the market elsewhere is more satisfactory than in Germany, to what extent should changes in the German rules be recommended, and whether new European rules are to be expected or should be proposed. The Ministry was anxious to be able to anticipate any need for new legislation or new policy development, and ensure that in the forthcoming Single European Market the German development industry was not disadvantaged by any lack of information about how development is undertaken in competitor countries.

It is impossible for one person or one team to gather all the information about land and property in other countries necessary for a well-founded comparison. The co-ordinators in Dortmund were glad and grateful to be able to fall back on the assistance of other members of PRODEST EUROPE, Property Development and Planning Studies in Europe, a consortium of several European universities which is seeking to identify the training requirements of participants in the European property development and planning fields and to promote research and the provision of an educational infrastructure for the development of professional skills in this field. Country reports were completed in the University of Newcastle upon Tyne for the UK, the University of Nijmegen for the Netherlands, the École Polytechnique in Paris for France and the University and Politechnic of Torino for Italy. Sweden is not yet a member of the EC and therefore not part of the EuProMa project, but a book for the series, adopting the same framework and structure, is being prepared in the Royal Institute of Technology, Stockholm. The invitation was extended to Sweden in view of the imminent agreement between the EC and EFTA to form the European Economic Area. There is the prospect of Swedish membership of the EC before long, as well as clear evidence that the Swedish property and real estate industry and profession is already operating on a European basis.

English was the working language of the different national teams and it was agreed that the national reports were to be written in English or (with the exception of the British report) in "Euro-English", the modern equivalent of the Latin of the Middle Ages. Therefore, English was also agreed to be the language of the publication. The task of series editing was shared between the project director and the UK members of the team, the latter bearing the much larger share of the burden.

To use the books prudently the reader should know a little about the methodological path of the whole project.

In order not to get lost among the many different specialist aspects of the task, each team operated on the basis of the same detailed examination pattern prepared by the co-ordinators in consultation with the other contributors. This pattern forms the basis of the chapter structure of each of the volumes in the series. Thus, comparability between the national studies is ensured.

The separation of land and property (Parts II and III of each book) may require some explanation. Although in most countries the great majority of the material is equally applicable to both, the two markets often do have their own operating characteristics. However, in discussion with the whole team when the framework was being formulated, it became clear that in the case of the Netherlands there were two quite different sets of procedures and actors, and that the best way to expose this on a comparative basis was by this separation in the examination pattern.

The operation of the land and the property market of any country is also determined by general economic, political and cultural conditions in the respective country, the constitutional and legal framework and the rôle and independence of local authorities. Factors such as forms of tenure, macroeconomic variables, performance of the economy, social changes, demographic development, owner-occupation, the requirements for land and the trends in spatial development had all therefore to be taken into account and form the subject matter of Part I of each book. The land market is being influenced, for example, by the conditions for and the process of land-use planning, the process of land assembly (including the development process), the process of construction and the regulations governing the first use of buildings. The level of detail to which the teams were required to work can be seen by looking at the following excerpt of the examination pattern for the framework for the urban land market.

Excerpt from examination pattern

SECTION II: THE URBAN LAND MARKET
1. *The framework within which the market of urban land functions*

1.1 *The legal environment*
     (The legal, esp. town-planning system as it affects the conversion of land to the first urban use.)

1.1.1 Law – Acts – Competences – Plans
   _   Hierarchy of competences
       (Hierarchie/Zuständigkeiten)
   –   Planning acts and their hierarchy
       (Gesetzeshierarchie)
   –   Plans (application, importance)
       (Pläne; Anwendung, Bedeutung)
   –   Presentation in a diagram
       (Darstellung in Schemata)
   –   Obligation/binding character of the plans
       (Verbindlichkeit der Pläne)
   –   Possibilities for higher tiers of administration to intervene
       (Eingriffsmöglichkeiten der höheren Ebenen)
(. . .)
   –   Environmental protection laws
       (Umweltrecht)
   –   Landscape protection (Landschaftsschutz)
   –   Water protection (Wasserschutz)
   –   Law of environmental impact assessment (EEC/85/337)
       (Umweltverträglichkeitsprüfung)
   –   Trade inspection/Nuisance
       (Gewerbeaufsicht/Immissionsschutz)
   –   Other relevant acts
       (air, noise, public traffic, housing, etc.)
       (Sonstige relevante Nebenrechte – Luft, Lärm, ÖPNV,
         Wohnungsbau, etc.)

1.1.2 Planning process for local plans
   –   Planning process (formal) for local plans, which confer the right to
       construct a building
       (lokale Ebene für Pläne mit Baurechtscharakter)

- Credit practice (credit securities, amortisation/repayment, interest rates
  (Kreditsystem: Sicherheiten, Zinsniveau, Tilgung)
- Restrictions on capital import
  (Beschränkungen für ausländisches Kapital)

1.2.2 Per sector
- Possibilities of financing
  (Finanzierungsmöglichkeiten)
- Investors (Investoren)

1.2.3 Land banking (Bodenvorratspolitik)

1.2.4 Transaction costs

1.3 *Tax and subsidy environment*

1.3.1 Taxes concerning the land market
- Taxation of land
  (Besteuerung von Grund und Boden)
- Taxation of other kinds of capital/assets
  (savings deposits, shares, capital income tax, . . .)
  (Besteuerung anderer Vermögensarten)
- VAT – value added tax (Mehrwertsteuer)

To understand and to evaluate the complicated land and property markets, three components have to be distinguished: (a) the framework within the market is functioning; (b) the process and interaction between different actors who, while operating within the same framework, have different aims and are playing certain specified rôles; (c) the outcome of the market, or the result of its operation. As a fourth and final step, the outcome, as well as the framework and the process leading to the outcome, all have to be evaluated.

Case studies were a central component of the project, six being provided from each country. These are essentially illustrations of the normal operations of the market. They are representative and not special cases, selected on the basis that they should illustrate different aspects of the functioning of the market and the development process in the respective country.

Clearly it was necessary when working on land and property markets to agree on criteria for judging the market, its performance and its outcome. We agreed to use as a basis for this the formula put forward by Hooper for the OECD Urban Affairs Programme:

> The general aim (of the land market) is usually to secure that development land is supplied in needed quantities, appropriate locations, appropriate tenure, at the right time and at appropriate prices, having regard to issues of economic efficiency and social equity. (Hooper, A., "Policy innovation and urban land markets", OECD Urban Affairs Programme, Paris 1989: 5)

It was for the national teams to use this formula to derive the criteria by which to judge the performance and the outcome of their land and property markets, including considerations of social equity and ecological aspects.

Having completed the national research reports, it seemed to all participants and contributors that the material assembled was too valuable to leave unpublished and only referred to in the final comparative study of the German team. The opportunity could not be missed to make these comprehensive studies about the land and property markets of major European countries available to a wider public.

As series editors we would like to thank all authors of the national reports and books for their support and co-operation in the publishing programme after they had thought they had finished all their hard work by completing the research reports. We would also like to thank the German Ministry for Regional Planning, Building and Town Planning for the generous agreement that the material assembled in reports commissioned by the Ministry and sponsored from its research budget can be disseminated to reach a wider audience in this way.

HARTMUT DIETERICH   RICHARD WILLIAMS   BARRY WOOD
Dortmund   Newcastle upon Tyne

SUMMER 1993

PREFACE

The report on which the French volume of the European Urban Land and Property Markets series is based was prepared by a team from ADEF, the Association pour le Développement des Études Foncières. ADEF is located in La Defense, Paris, and is a non-profit association of property professionals, developers, public authorities and academic researchers formed to exchange and publish information and research on all questions relating to land policy through its own journal, Études Foncières, and through books and reports.

The assistance of the following in compiling this study is acknowledged: Ingrid Ernst for the Alsace case study; Alain Motte, formerly of the Institut d'Urbanisme de Grenoble, Université Grenoble II, now of the Université Aix en Provence, for the Meylan case study; François Paillé for the Citroën case study; and Stephane Muzika for the Lorraine case study. Also, for their contribution to the study of the urban property market, Bruno Lefebvre, Yannick Martin, Claire Ombrouck and Sylvie Occhipinti, economist at CEREVE, Université Paris X at Nanterre. Finally, for assistance in the preparation of the text for publication, Charlotte Andrews and Sharon Holt.

We wish to record our thanks to all the other collaborators in the project, but of course we remain responsible as authors for the contents of this book.

RODRIGO ACOSTA   VINCENT RENARD
                        PARIS, MARCH 1993

ACKNOWLEDGEMENTS

The assistance of the following in compiling this study is acknowledged: Ingrid Ernst, Alain Motte, François Paille, Stephane Muzika (case studies); Bruno Lefebre, Yannick Martin, Claire Ombrouck, Sylvie Occhipinti (the urban property market); Charlotte Andrews, Sharon Holt (editing assistance, Newcastle).

ABBREVIATIONS AND ACRONYMS

AFTRP	Agence Foncière et Technique de la Région Parisienne (land agency for the Paris region)
AFU	Association Foncière Urbaine (urban land pooling association)
ANAH	Agence Nationale pour l'Amélioration de l'Habitat (national agency for housing renewal)
APL	Aide Personalisée au Logement (personal housing allowance, used for new housing schemes with low interest rates)
AUDIAR	Agence d'Urbanisme et de Développement Intercommunal de l'Agglomération Rennaise (see Ch. 6.1)
bln	billion
CAECL	Caisse d'Aide à l'Équipement des Collectivitées Locales (national fund for local authorities)
CAHR	Comité d'action du Haut-Rhin (see Ch. 6.3)
CDC	Caisse des Dépôts et Consignations
CFF	Crédit Foncier de France (state bank for funding public works and housing)
CGI	Code General des Impôts (general fiscal code)
CLF	Crédit Local de France (bank for local authority loans)
CNRS	Centre National de la Recherche Scientifique
CODESPAR	Comité du Développement Économique et Social du Pays Rennes (see Ch. 6.1)
COS	(Coefficient d'occupation des sols) floor area ratio (FAR)
CREDOC	Centre de recherche pour l'étude et l'observation des conditions de vie
DATAR	Délégation à l'Aménagement du Territoire et à l'Action Régionale (Government agency for regional policy)
DDE	Direction Départementale de l'Équipment (Government Department for urban affairs)
DIA	Déclaration d'intention d'aliéner (declaration of intention to sell)
DPU	Droit de préemption urbaine (urban preemption right)
DUAR	District Urbain de l'Agglomeration Rennaise (Rennes district, 28 municipalities)
DUP	Déclaration d'utilité publique (notice of public interest)
EAP	Economically Active Population
EC	European Community
EPBS	Etablissement Public de la Basse-Seine (public land bank in Normandy)
EPML	Etablissement Public de la Métropôle Lorraine (public land bank in the Lorraine Region)
ERDF	European Regional Development Fund (of EC)
ERM	Exchange Rate Mechanism (of EC)
FAR	Floor Area Ratio (see COS)
FAU	Fonds d'aménagement Urbain (urban management fund for urban development schemes)
FDES	Fonds de Développement Economique et Social (economic and social development fund for urban authorities)
FEDER	see ERDF
FF	French franc

FIAT	Fonds Interministeriel d'Aménagement du Territoire (national fund for regional policy)
FNAFU	Fonds National d'Aménagement Foncier Urbain (national fund for urban land development)
FNAIM	Fédération nationale des Agents Immobiliers (National real estate federation)
FNPC	Fédération Nationale des Promoteurs Constructeurs
FSGT	Fonds Spécial des Grands Travaux (Special fund for major recycling projects)
GDI	gross disposable income
GDP	gross domestic product
GNP	gross national product
ha	hectare
HLM	habitat à loyer modéré (social housing with controlled rent)
IMO	Ministry of Finance databank
INSEE	Institut National de la Statistique et des Études Économiques
LDC	legal density ceiling (see PLD)
LOF	Loi d'orientation foncière (town and country planning act of 1967)
LOV	Loi d'orientation pour la ville (1991 Land Act)
m	million
NA	non aménagée "A" (future urban land in POS)
NB	ordinary natural areas "B" in POS
NC	agricultural areas "NC" (where building forbidden in POS)
ND	conservation areas
OCDE	see OECD
OECD	Organisation for Economic Cooperation and Development
PAE	programme d'aménagement d'ensemble (special infrastructure financing scheme and special exaction area)
PAP	prêts à l'accession à la propriété (subsidized loan scheme for house purchase)
PAT	prime d'aménagement du territoire (public funding for land reconversion and management schemes)
PAZ	plan d'aménagement de zone (zone development plan)
PC	Prêts Conventionnés (subsidized loan scheme for house purchase)
PLA	Prêt locatif aide (cheap loan for social housing development)
PLD	plafond legal de densité (legal density ceiling)
PLI	Prêts Locatifs intermédiaires
POS	plan d'occupation des sols (local land-use plan)
RATP	Régie Autonome des Transports Parisiens (Paris regional transportation authority)
RNU	Régles Nationales d'Urbanisme (national regulations for control of development where no POS exists)
SADI	Société d'Aménagement et de Développement de l'Isère
SAFER	Agency for land management and rural organisation
SCI	Société civile immobilière (private property company)
SD	see SDAU
SDAU	Schéma Directeur d'Aménagement et d'Urbanisme (district local plan)
SEM	Société d'Economie Mixte (Public-private development company)
SNAL	Syndicat National des Aménageurs-Lotisseurs (professional developers' trade union)
SNCF	Société Nationale de Chemins de Fer Français (French national railway company)
TGV	Train à Grande Vitesse (high-speed train)
TLE	Taxe locale d'équipement (local development tax)
TSE	Taxe spéciale d'équipement (special Savoie Olympic tax)

U   urban zones (in POS)
UDC   urban development corporation (UK)
VAT/TVA value added tax
ZAC   Zone d'Aménagement Concerté (comprehensive development area)
ZAD   Zone d'Aménagement Différé (deferred development zone)
ZH   Zone d'Habitation (housing zone)
ZI   Zone Industrielle (industrial zone)
ZIF   Zone d'Intervention Foncière (land intervention zone)
ZUP   Zone à Urbaniser en priorité (priority zones for urban development)

PART I
Overview

CHAPTER 1

Basic information

1.1 Constitutional and legal framework and institutions

The constitution and the French territorial system

France is a unified non-federal republic and has had a long tradition of centralism, with uniform law applied to the whole country. In 1958, under General Charles de Gaulle's government, a new constitution was adopted that established a balance of power favouring the executive branch (the president and his cabinet). By contrast, the legislative body, the *Chambre des Députés* (Chamber of Deputies) of the National Assembly, which consists of 577 members elected for five years, was granted fewer powers (Fig. 1.1). The French president is elected directly for a seven-year term and is free to appoint the government. The president has various means to restrict legislative dissent: his prime minister can force the passage of laws without debate (Art. 49-3) or under certain conditions can dissolve the chamber of deputies, thereby calling a general election. Even though there has been a trend towards decentralization, government is still centralized, with policy-making remaining in Paris. The main ministries (for example, urban affairs and finance) are located in Paris; other external offices at the departmental (e.g. *Direction Départementale d'Équipement* (DDE) or ministry of urban affairs) and regional (*Délégation à l'Aménagement du Territoire et à l'Action Régionale* – DATAR) levels are found in the other major cities.

Before 1983, the French central government had specific means whereby it set up guidelines for urban policy and implemented comprehensive policies (e.g. urban reconstruction after 1946 and industrial planning). Before 1972, territorial organization was based on two key levels of sub-national jurisdiction: *communes* (municipalities) and *départements* (departments). Over 36,000 town councils or municipalities administered locally and 99 departmental bodies (95 in mainland France and 4 in its overseas territories) had wider regional power. The 1958 constitution stated explicitly that the three levels of territorial organization – the central state, the departments and the municipalities – should share territorial powers.

2

However, central state officials, notably the *préfets* (prefects), were empowered to act at the local level. Prefects supervised mayors and municipal assemblies, and served as an administrative link to the central government, which operated as an important tier in territorial government.

The French territorial system after 1983

On 2 March 1982, President Mitterrand's government passed decentralization legislation (enacted in 1983) that altered significantly the power relationships between the different territorial levels. Briefly, the previous three levels of administration were increased to four: the municipality, the department, the region and the central state.

At the local level, the 1980s were characterized by a trend towards municipal self-administration (Fig. 1.1). Municipal bodies progressively acquired wider prerogatives and broader powers on various matters, including fiscal

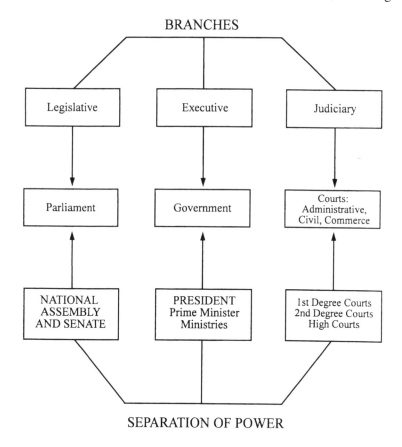

Figure 1.1 Political system.

issues (the levying of local taxes), town-planning, building permits, urban transportation, primary schools and local roads. The large number of municipalities causes serious problems: the gap between dynamic and less dynamic municipalities has resulted in a spatial disequilibrium. Some have become poorer and others richer in terms of taxes, infrastructure and demography. Of 36,000 municipalities, 28,000 have fewer than 1,000 inhabitants.Moreover, intercommunal restructuring has not solved these problems, and to counteract them local government legislation, known as the Joxe–Marchand statute, and laws on *solidarité communale* (municipal cross-subsidization) were passed in 1991.

The decentralization trend from 1983 has greatly modified departmental organization. There is an ambiguity between the executive and the legislative bodies. Although a department's citizens directly elect its *conseil général* (general council), its president is chosen by his/her peers from among their number, to become the holder of departmental executive power. These executive officials have gradually replaced prefectural authority on urban matters. Departmental administration has jurisdiction over non-urban issues, school transportation, secondary school education and social welfare. The prefects' rôle has broadened to that of overseeing the implementation of national policy at lower territorial levels, and their only remaining powers arise out of their duty as state representatives to serve as a channels of communication and as mediators.

Decentralization has also strengthened regional administration (Fig. 1.1). The elected council of a region chooses a president who holds executive power on behalf of the region. Broadly, a transfer of power has occurred from regional prefects to regional executives in regional infrastructure (e.g. airports and waterways), regional planning and environmental protection.

Today, the central state's rôle in local affairs is to provide administrative services and set up policy guidelines rather than to implement policy. The decentralization period has introduced new urban planning methods: Before 1983, the central state exercised *ex ante* control over territorial organizations. Today, its rôle is to control *a posteriori*. For example, administrative courts may consider the legality of measures taken by local administrations and cases may be taken to appeal in the Conseil d'Etat (supreme administrative court). And in each region a *Cour Régional des Comptes* (regional audit court) reviews financial and fiscal procedures occurring at all administrative levels.

Politicians are currently debating the powers of these territorial levels. In respect of the transfer of powers, the trend could change in the future under the terms of the Joxe–Marchand Statute, which will allow the prefects a more decisive rôle. A further consideration is whether the four levels of government should be restructured; some commentators advocate the abolition of the departmental level and some argue that even the number of regions could be reduced, from 22 to 10 or 12, to enable France to be more competitive within the Single European Market.

Laws relating to land-use and tenure

The birth of the Fifth Republic (1958), and the 1804 *Code Civil* (civil code) are the legal bases for the French constitution and the laws that spring from it. The 1958 Constitution provides for the two technical sources of French law: *lois* and *décrets* (statutes and decrees). These differ both in origin and scope. Statutes are limited to certain subject areas outlined by the constitution, including general principles of ownership rights, town planning, and land policy, and must be passed by a majority in both parliamentary chambers (national assembly and senate).

Broadly speaking, in France there is an abyss between, on the one hand, a constitutional emphasis on the protection of private land and property rights and, on the other, a growing body of regulations that constrain this legal guarantee in favour of the public interest. In theory, the right of ownership is considered inalienable and sacred (as defined by the 1804 civil code). Nevertheless, the notion of public interest – in other words, collective goals – often subordinates this precious right (ADEF 1990).

Evoking the 1958 Constitution and referring to the "Declaration of Human Rights" no longer provide grounds for legal claims, except when a statute law is referred to the *Conseil Constitutionnel* (constitutional council), which is empowered to determine its constitutionality. For example, a general statement of the pre-emption right in a 1985 statute was referred to the constitutional council, which adjudged that it was compatible with the right of property as the constitution defines it. Once the *Journal Officiel* (official bulletin) publishes a statute or regulation, its legality and its consistency with the constitution can no longer be challenged. Only another statute or regulation may alter or annul previous laws.

The judicial courts

Another component of the French legal system is the judiciary. Two types of *tribunaux administratifs* (tribunals or civil courts) – first and second degree – have different and unrelated tasks and rôles. If conflict arises between two parties, they go to a first-degree civil court appropriate to the subject matter and with appropriate powers, for example, the *Tribunal de Commerce* (commercial court) or the *Tribunal d'Instance* (criminal court). The appropriate first-degree court settles the issue according to the facts presented and the law.

It is necessary to draw attention to the courts' ambiguous position as far as their independence is concerned. Although the eight administrative tribunals are technically part of the government, since they are nominated by the justice ministry, these tribunals are viewed as independent and non-partisan entities. Nevertheless, political interference may occur. The system of judicial courts, as well as the statutes and regulations influences the application of the different legal codes. The latter are organized by topics or issues such as mining, forestry, environment, urban planning, etc.

Forms of tenure and inheritance rights

A main principle of inheritance rights is equal shares for heirs. This results in high levels of plot subdivision and therefore complicated land–ownership patterns. This is especially true in urban areas where there are approximately 90 million individual plots of land, creating difficulties and costs for urban development. In urban areas, land is mostly in full freehold ownership. Leasehold is widespread for agricultural land but unusual in urban areas.

1.2 The economic framework and recent changes

Recent developments in the economy

It is clear that the present-day importance of public-sector intervention in France is the consequence of two major periods of policy-making. There was a significant period of nationalization just after the Second World War which incorporated plans to upgrade French industry. De Gaulle's government nationalized electricity supply, the railways, the Paris transportation system (RATP), four major banks, major insurance companies, Air France and Renault. Economic crises in 1973 and 1979 also led to a restructuring of the French economic apparatus. In the 1981–2 period, the first Socialist government enlarged public-sector ownership, adding more banks and major industrial firms such as Pechiney, Thomson and Rhône-Poulenc.

Although state involvement in the economy as a whole has been strong, through such measures as industrial policy, price controls and active monetary policy, the rôle of the private sector has grown. Its rôle and influence have affected the economy regardless of the political party in power. Ironically, the Socialist government of 1983-6 slowed interventionism.

The arrival of a conservative government (1986–8) reinforced this trend, with most of the 1982 nationalizations being privatized. However, the policy of privatization and deregulation was not carried out to its fullest extent because of political changes associated with the presidential elections of 1988. Since then, the status quo has remained.

The private sector has invigorated the economy (a major example being the 1990 Volvo-Renault agreement), and joint ventures seem to be on the

6

increase. By the end of the 1980s, the economy was performing healthily, as measured by inflation, GDP, etc. Table 1.1 examines French economic performance in terms of productivity and investment, comparing it with the economic performances of 6 other OECD countries.

Table 1.1 Productivity and investment.

	Productivity growth p.a. (1979–88)		Investment/GNP p.a. (1979–87)	
Canada	1.2	7	15.1	3
France	2.0	2	14.3	5
W Germany	1.6	5	14.8	4
Italy	1.9	3	15.1	2
Japan	3.0	1	23.3	1
Sweden	1.4	6	13.7	6
UK	1.8	4	13.0	7
USA	1.0	8	12.7	8

Source: OECD and INSEE.

France's second position in Table 1.1 for productivity indicates continued strength. But its fifth position for investment as a percentage of GNP reveals its weakening position. The *Institut National des Statistiques et des Etudes Economiques* (INSEE) collects census data and carries out sector studies within the French economy, examining seven separate economic sectors. Throughout this study, five of these INSEE economic sectors will be used. In general all economic data quoted here essentially comes from two INSEE sources: its national accounts and social data (INSEE 1989, 1990). Information from relevant ministries has also been used.

Gross domestic product (GDP)

Economic growth was faster in the 1950–73 period than in any other period in recent French history. As Figure 1.2 shows, during the 1960s average annual GDP growth rates were equal to or higher than 4.4 per cent. Since then, the French growth rate has evolved along lines similar to those of other industrialized countries. The late 1970s and early 1980s were marked by a phenomenon known as stagflation, the co-existence of high inflation rates and economic recession or stagnation. From 1987 to 1992, economic growth became stronger.

Inflation rates

Throughout the 1970s and through to the mid-1980s, recurrent high levels of inflation were a major issue in economic policy. Between 1963 and 1969 the average annual rate remained low at only 3.8 per cent per year. By contrast, after 1973 the price index averaged 10.7 per cent annually, higher

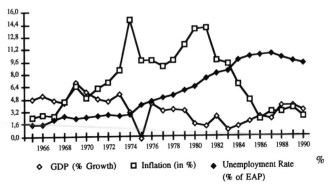

Figure 1.2 Major economic indicators.

(by 1.7 per cent) than the average OECD annual rate. The government's objective was therefore to curtail inflation.

After a short breathing space (1974–8), high annual inflation rates of over 9 per cent again undermined government policy. By 1985 annual inflation had been reduced to under 3 per cent and it has since oscillated around this level. By 1990, the government had attained its policy objective of stable inflation.

The unemployment rate

A combination of structural and temporal factors has generally contributed to the growth of the active population since the 1960s. From 1975 to 1989 it grew from 22 to 24 million. INSEE forecasts that at the turn of the 21st century it will reach more than 26 million (INSEE 1989).

Unemployment affects certain social groups disproportionately: young people, women and the less well educated.

The 1981–4 economic recession exacerbated unemployment. Employment declined in construction, industry, and even the service-related sectors at an annual average rate of 1.5 per cent until 1984. Unemployment has persisted because of structural factors, the rigidity of the labour market and a mediocre economic situation. Figure 1.2 clearly illustrates the trend towards a higher unemployment rate over the past 20 years. In mid 1993, 2.9 million people were unemployed, and unemployment remains a key political and economic issue.

The balance of payments

Another key economic issue is the balance of payments. Throughout the period 1982–9, France's combined deficit increased, and only once (in 1986) was a surplus recorded (Fig. 1.3). Although a positive balance has been maintained in the tertiary sector, the deficit problem is recurrent.

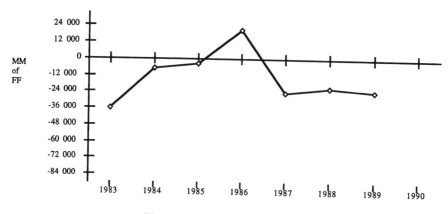

Figure 1.3 Balance of payments.

Investment

Investment levels, as indicated by the ratio of investment to GDP in the past two decades, showed a relative decline up to 1985 (Fig. 1.4). Along with the economic upturn, private and public investments have subsequently grown: in 1985, the global investment rate showed a 2.1 per cent increment in volume, with 3.8 per cent growth rates in both 1986 and 1987.

Figure 1.4 covers five major sectors of the economy. If these are disaggregated it can be observed that:

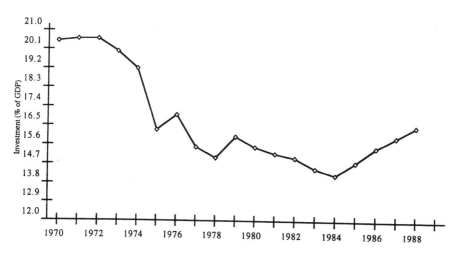

Figure 1.4 Investment.

9

○ large state-owned enterprises have reduced their investment levels on equipment by 7 per cent reduction (in volume) since 1985;
○ the agricultural sector has followed the same trend, with a 14 per cent reduction since 1985,
○ investment in competitive industry continued to grow slowly, with a growth rate of little more than 1.5 per cent in the three years from 1985 to 1988;
○ investment in the commerce and services sector was very erratic because of its dependence on household consumption, which makes it very sensitive to the economic situation (+1 per cent in 1985, +11.3 per cent in 1986, +6 per cent in 1987),
○ the construction and infrastructure sector investment rates fluctuated widely in line with sales of houses and flats, which in turn depended on interest rates and government policy on tax incentives and credits (−5.8 per cent in 1985; +11.3 per cent in 1986; +3 per cent in 1987).

Income levels

Based on 1980 constant values (FF), GDP per capita has increased at a constant pace (Fig. 1.5) having expanded from FF39,961 to FF52,121 in one decade (1970–80). In 1989, it hovered at around FF61,000 (INSEE 1990). After three economic cycles, gross disposable income (GDI), which had been fluctuating between 10 per cent and 17 per cent, slid to very low levels after 1982. It bottomed out in 1986 (5.3 per cent) and 1987 (3.8 per cent). The

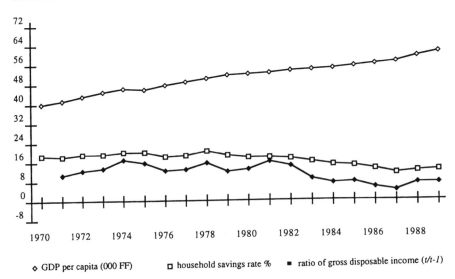

Figure 1.5 Household indicators.

10

income level, measured by GDI, has decreased during the last few years of recession, especially between 1982 and 1987. Nonetheless, it has now rebounded slightly because of a slight recovery.

Household savings rates

According to a recent study on household savings, the average family savings rate remains insufficient (L'Hardy 1990). In 1989, it was 12.3 per cent of disposable income (income remaining after taxes and social security charges). This rate has been stable since 1986 (Fig. 1.5). Nevertheless, the household saving rate in the 1970s and early 1980s was much higher than now. In 1970 it reached 18.7 per cent, and in 1975 20.2 per cent. By contrast, it was 17.6 per cent in 1980. With the exception of Germany and Japan, the same tendency in household savings rates has been observed in other industrialized countries.

Government debt

Public debt has risen significantly since 1976. The French government has increased its debts directly (bond issues) or indirectly (by financing state-owned companies' debts). A recurrent dilemma for it is whether to borrow to finance public expenditure. Another noticeable phenomenon has been the indebtedness of municipal and regional bodies. Since decentralization started they have increasingly either borrowed directly from banks and other financial institutions or they have guaranteed other public or private–public administrations such as *sociétés d'economie mixte* (SEMs – public–private development companies) or football clubs. Financial partners, of course, encourage and favour this trend and support local authorities' autonomy in handling their investments and capital assets.

Capital mobility

The French have a long tradition of holding onto their money rather than investing it in the financial markets. In the past, according to a thorough study of French wealth (Babeau 1988), land assets and property (*la pierre*) have been favoured investments, but this is changing in line with evolving social values (see Ch. 3.3). Today the average adult invests more in bonds, securities and stocks than was the case in previous decades (INSEE 1989).

Furthermore, different patterns of investment in financial assets according to age group have been found for the past decade (CREDOC 1990). Three age groups: 34–49, 50–64 and 65+ have since mid-1985 surpassed the 15 per cent threshold of financial investment in relation to their total investment. People over 50 have investment rates of around 22 per cent. By contrast, the younger age group, 21–34, had not yet reached the 15 per cent threshold in 1989.

Table 1.2 Distribution of principal assets.

Year	Real estate					Financial assets[4]	
	Investment asset[1]	Primary residence[2]	Secondary residence	Other estates	Land assets[3]	Saving or housing acc	Life insurance
1978	7.9	46.7	9.4	5.5	15.9	66.4	QNA
1979	8.7	51.7	8	7.8	15	67.5	QNA
1980	8.4	47.1	9.1	8.6	18.3	67.6	QNA
1981	9.4	47.4	10	7.6	15.5	68.1	QNA
1982	9.1	46.8	8.3	7.9	15	66.6	QNA
1983	10.9	46.4	9	9.1	15.3	68.9	QNA
1984	9.9	50.9	7.9	7.8	12.8	68.9	QNA
1985	13.2	51.3	11.6	9.2	14.2	69.9	26.2
1986	15.6	49.9	9.8	8.2	12.8	68	29.5
1987	19.1	54.5	8.9	9	14.8	63.4	35.2

Source: CREDOC. The figures are percentages of a representative national survey of 2,000 persons (total can be equal to more than 100% due the multiple-choice method used).
Notes: 1: Bonds and stocks. 2: Old owners or recent ones. 3: Lots or forests. 4: At least holding a S&L account or a Housings Savings Plan.

At the beginning of 1990, according to another representative household survey (CREDOC 1989), 19 per cent of adults declared ownership of financial investments (securities, stocks and shares). Even though this is one point below the figure for the previous year (probably as a consequence of the 1987 stock market crash), it has more than doubled in 10 years from the 8.4 per cent recorded in 1980. This trend started long before the privatization period of 1986–7. Because of high added values and profitable returns in the short term, these investments are attractive to households. In brief, shares and securities have begun to replace the holding of cash and other traditional methods of saving.

Interest rates

From 1979 to 1982, interest rates increased dramatically to over 15 per cent. Later, they became stable and even decreased to under 10 per cent. In this respect, French policy followed the major trends of the industrialized countries' monetary systems and in particular that of the German Bundesbank. During the 1960s and 1970s it was profitable to acquire and possess land and real-estate assets because of low real interest rates. However, a reversal occurred during the 1977–8 period, with lower inflation causing an increase in real interest rates. This had a direct impact on land and property markets, especially on land-banking. During the 1980s, real interest rates grew steeply, modifying the behaviour of developers, builders and property investors because of the high opportunity cost of land-banking (see Ch. 3.2).

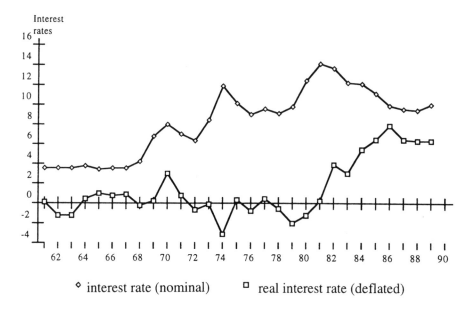

◇ interest rate (nominal)   ▫ real interest rate (deflated)

Figure 1.6 Real and nominal interest rates.

Forecasts for 1993 and the following years

Before exploring some economic projections, let us first consider an inherent element of French society, the *"maladie imaginaire française"* or French hypochondria. A common thought reflecting the state of French society, economy or both can be globally summarized along the line of: "we, the French, are far behind such and such country in this and that field." While in some respects this may be true, France remains and will no doubt continue to be a major economic actor on the international scene and particularly in Europe.

In order to forecast how the French economy will fare during the 1990s, one should consider recent political events such as the Gulf War and the demise of the communist regimes of central and eastern Europe, and the economic implications of such events. The economy is currently going through a transition period. An evident slowdown in most sectors has affected growth of GDP. Forecasters face a serious dilemma: is the economy at the end of a short cycle (the boom period 1987–9) or at the beginning of a readjustment period? The latter hypothesis suggested a durable boom period from 1987 onwards.

At the end of 1991, it was possible to argue that an economic upturn might reverse the economic slowdown. Some positive signs appeared in the economy: lower inflation and an upswing in terms of foreign trade. How-

ever, certain counter-indicators have arisen: an increase in the unemployment rate (the February 1992 figures were the worst since 1984) and the
possibility of a decline in the GDI of households. INSEE forecasts a mild recovery at the end of 1992, with estimates of from 1 per cent to 2 per cent
growth in GDP. In spite of that, a recession is developing in the first sem-
ester of 1993, and the GDP forecasts are around −1 per cent for the year
1993. France seems to be going through the current economic crisis with
some difficulties, the two major problems being the unemployment rate and
international economic imbalances.

Prominent factors include the economic recession in the USA (although
there are some positive signs there), the monetary policy of other European
economic partners, especially Germany, and the eventual consequences of
the inception of the Single European Market in 1993. Rises in interest rates
in other European countries (especially in Germany because of inflationary
tension caused by reunification) could have affected French interest rates and
eventually disrupt monetary policy. The government therefore hopes to
stabilize exchange rates within the European Community and reinforce the
exchange rate mechanism (ERM).

After the 1990–1 slowdown, it was hoped by all economic forecasters that
economic growth could restart and continue into 1993 at a rate of 3 per cent
growth of GDP. It is predicted that this will then diminish slightly to around
2.5 per cent in the following years (1994–7). Meanwhile, major international
forecasters such as the OECD remained sceptical about French GDP growth.
They suggested that the current level of GDP would diminish after 1991, as
it has.

Moreover, current political issues in Europe surrounding the implementation of the Maastricht Treaty and the construction of European Monetary Union are expected to affect the French economy, though the consequences are not clear.

1.3 The social context

Demographic and household structure: demographic and societal changes
Recent demographic and societal changes are slowly reshaping family and
household structure. The last available census figures (1990) indicate that
French population growth is among the most dynamic in the European
Community, ranking fourth, even though the fertility rate has remained low
at 1.8 per female.

In 1990 the population was 58,073,553 (56,614,493 if overseas departments are excluded). Population has grown at an average rate of 0.5 per
cent since the 1982 census. This is almost the same as in the period 1975–

82. The natural surplus (births minus deaths) accounts for four-fifths of this 0.5 per cent growth.

Social and demographic changes took place between 1975 and 1990. While the fertility rate remained at about 1.8 children per female, the number of births out of wedlock rose. Moreover, the number of marriages has declined relatively (even though there were indications that this trend slowed after 1988) and there was a continuing decline in the mortality rate, with on average one year added to life expectancy every three or four years (INSEE 1990: 18). The result, as in other countries, is that the population is ageing, with an average life expectancy of nearly 70 years.

Single-parent families are a growing feature of the past decade. An increasing number of divorces and unmarried mothers means that there are fewer large families than before. In turn, this has contributed to a household explosion. The average number of persons per household has decreased from 3.1 in 1962 to 2.7 in 1982 and 2.53 in 1990 (INSEE 1990: 284). Furthermore, a larger number of households consist of one person only (26.9 per cent of total households in 1990). This is expected to grow to 27.5 per cent by 1995. This category of household, consisting, for example, of unmarried young professionals and the elderly, is the category most likely to increase.

Effects on housing demand to 1995

New family and household structures affect housing demand. There are certain housing demands that cannot be met today. The mobility of certain groups, such as high-ranking professionals, students and freelance workers, has led to a higher demand for housing. Unmarried couples and divorced persons also contribute to the demand for temporary accommodation. The result is a different structure of housing demand.

This should change the nature of housing policy, which envisages housing demand as characteristically consisting of traditional couples with children seeking single-family housing. There is a strong demand for rented accommodation and consequently more construction. Some observers have estimated that between 1987 and 1989 there was a need for 349,000 housing starts per year: 239,000 for principal residences in order to provide new habitat, 67,000 for stock renewal and 26,000 to respond to the demand for second residences (Table 1.3).

More importantly, an increase in household numbers will accelerate the demand for housing at an annual average rate of 329,000 units between 1987 and 1995. Future generations will continue to arrive on the housing market in large numbers, while the aged population rises. Corollary effects will follow. Diversified family structure and a slower life-cycle evolution (because of studies, active life, marriage, etc.) will have a dual impact: a higher residential mobility and a stronger demand for rented housing, especially in town centres (Bonvalet, in INSEE 1990: 285). What is more, demand coming specifically from young urban professionals and the elderly will continue to prime the pump of housing starts.

Table 1.3 Housing needs 1987–95 (000s).

	1987–9	1990–2	1993–5	Annual average flow 1987–95
Principal residences including	**286**	**275**	**245**	**269**
Dwelling demand due to	239	228	198	222
Demographic need	171	176	156	168
Single person need	68	52	42	54
Stock renewal	47	47	47	47
Empty housing including	**17**	**6**	**16**	**13**
Parc flow	22	21	18	20
Parc renewal[1]	(5)	(5)	(5)	(5)
Secondary residences including	**46**	**46**	**43**	**45**
Growth of Parc	26	26	23	25
Stock renewal	20	20	20	20
TOTAL NEED	**349**	**337**	**301**	**329**

Source: INSEE Social Data 1990. *Note:* 1. A negative figure is due to all the flows that affect the existing stock.

Demand will nevertheless continue to reflect market conditions of costs, supply and interest rates. Regional discrepancies and inter-urban differences will influence and reinforce inherent demand structures. The case of the Paris region can be taken to illustrate this point: new housing starts average nearly 35,000 per year whereas current need is estimated to be 75,000. Therefore an annual deficit of 40,000 starts should be added to the existing deficit in the Ile-de-France.

Housing needs are not being fully met and the consequent structural deficit is worsening. This deficit may be even more difficult to reduce if account is taken of the immigrant population's housing needs.

Immigrants and their status

Passion has overwhelmed reason more than once where immigration is concerned. One extreme argument can challenge another, depending on the ideological point of reference and other parameters. However, certain facts stand out: French policy has encouraged immigration because of the needs of the economy and certain geopolitical factors. Globally, the immigrant population as a percentage of total population has remained lower than 10 per cent, growing from 4.38 per cent in 1946 to 8 per cent in 1990.

Between the wars, the foreign population consisted mainly of immigrants from neighbouring European states: Italians, Poles, Spaniards and some Belgians made up the bulk of the 2 million immigrants (INSEE 1990: 312). The government implemented an open-door policy, even though there were some outbursts of xenophobia during socioeconomic crises.

The fundamentally open-door immigration policy continued in the postwar era and was a major characteristic of policy until the 1970s. Immigration patterns in the 1950s and 1960s differed from prewar ones: most of the new labour force came from such areas as North and sub-Saharan Africa, Asia and Latin America. Most industries encouraged this open-door policy so that they could hire cheap labour and break labour-union monopolies. In response to uncontrolled immigration in the 1970s, stiffer regulations were introduced. Thus, on 3 July 1974, a decision of the Conseil d'Etat temporarily stopped foreign labour importation, invoking higher unemployment levels and recession as justification.

The depth of the economic crisis of the 1970s resulted in the creation of more stringent laws that modified and completely interrupted official immigration. Generally today the exceptions are those seeking political asylum and European Community nationals. The flow of immigrants, however, has not been stopped. It is estimated that 60,000 people have entered French territory annually since 1974. Recent figures show that 4,500,000 foreigners live in France.

Today, negotiations are under way to harmonize immigration policy at the European Community level under the Schengen agreements, signed in 1990 and passed by the French government and approved by parliament in 1991. French policy-makers are working with other European Community on the convergence of member-state policies whereby each country should put aside its geopolitical and economic prerogatives in order to solve the problem at a European level. The issue is linked with other urban and economic problems such as lack of urban infrastructure, high unemployment rates and social degeneration in ghetto areas. For instance, the immigration problem exacerbates the housing problem because most immigrants live in the *banlieues* (peripheral areas of towns and cities), where economic and social problems are concentrated. To respond to this acute problem, a law has recently been passed (*Loi d'orientation pour la ville*, July 1991; see Ch. 2).

Geographical mobility

Even though most workers have little propensity to move for professional reasons, geographical mobility depends today on economic factors. According to an INSEE survey, certain individuals move with increasing frequency and over ever greater distances than others. Thus, local migration leads to migration between regions. An individual who had moved during the period 1968–75 was five times more likely to move in the next period, 1975–82, than someone who had not moved during the earlier period. Typically, civil servants and workers employed by government agencies moved more frequently and over greater distances than the average (see Part III). Sixteen per cent of the population of continental France had lived in different regions at the two different dates of 1962 and 1982. During the 1970s, an exodus from the Paris region was observed, as was an attraction towards the western and southern regions.

Geographical migration reinforces employment turnover. During the past 15 years, the nature of labour-force turnover has changed somewhat. In the period 1965–70, many exchanges took place between economic sectors. Industry served as the platform for these exchanges because of its vitality. However, between 1980 and 1985 industry lost ground, and consequently unemployment became the key to such turnovers in absolute and relative terms, and the public, construction and civil engineering sectors competed more effectively for labour. Figures for 1986–8 show that male workers form the core group of mobile professionals, and that female workers move less. To explain these evolving mobility rates and geographical migrations, one needs to explore not only the rigidity of the housing structure, which may not allow or facilitate mobility, but also changes in social values.

Mentality and social changes in society

The French mentality throughout the 1980s has evolved around four themes: relative standardization of social and political views, money, stress and individualism. The *Centre de recherche pour l'étude et l'observation des conditions de vie* (CREDOC) has operated as an observatory of such social changes and traits throughout the 1980s.

Perspectives in society on various issues ranging from marriage to progressive changes in economic policy and society have become more standardized. The long-standing antagonism between Paris and the provinces has receded, Parisians are more pragmatic on many issues today and in certain respects they seem to have become politically conservative. Meanwhile, provincial people, who were relatively more conservative in the past, appear to be catching up with Parisian perspectives. The result is convergence towards uniform ways of perceiving and doing things, e.g. consumption habits and frequent travelling. Television is a key factor in this and it has become pervasive and hegemonic in society: in 1990, 75 per cent of people watched it daily, whereas only 50 per cent did so in 1980.

One of the fundamentals of social evolution in the 1980s was a changing attitude towards money. Although the high-finance world of the stock market and the wave of the golden boys and yuppies was criticized in the 1980s, there has been a major change in attitudes. Greater acceptance of the values of the business world and entrepreneurial tendencies now exists and has had a clear consequence: the French are less reticent about high salaries. In other words highly paid professionals are no longer criticized as they were in 1980. For example, in a sample survey, the percentage of respondents who considered that a firm's chairman or managing director was paid too much money dropped from 54 per cent in 1978 to 40 per cent in 1988.

At the same time, a growing gap between the "haves" and the "have nots" has appeared. Two reasons for this can be stated here: changing attitudes to the moral acceptability of wealth accumulation, and higher investments in securities and stocks rather than in real-estate. In parallel, a wave of rugged individualism is breaking out in society, as can be witnessed in other west European countries.

The car, for instance, is considered a sign of independence and individualism. Because of the inherent environmental problems of cars, the French were strongly opposed to their use in 1981 (37 per cent). But this declined to 22 per cent in 1988. All in all, a retreat from the civic spirit that tried to rationalize car use has occurred. More and more people disagree (24 per cent in 1988) with the statement "car use should be limited in order to improve urban traffic."

Alongside the unemployment problem, health troubles have appeared that are psychological in origin: the proportion of people suffering from back pains (45 per cent in 1988 and 1989), nervous breakdown (44 per cent) or insomnia (28 per cent) has nearly doubled in a decade. The percentage of people suffering from headaches has gone up to 20 per cent. In short, stress and competition go together.

1.4 Land and property markets and the building industry

The owner-occupancy rate

Even though each country has its own perceptions of family and society, statisticians have tried to compare definitions and produce data that reflect owner-occupancy patterns in industrialized countries (INSEE 1990: 285). One major trend stands out: the number of owner-occupiers in European Community countries has steadily increased (Table 1.4).

France's evolution in particular vividly illustrates this pattern. As the number of households increased by 40 per cent in the 25-year period 1963–88, the proportion of owner-occupiers grew from 42.2 per cent to 54.3 per cent of all occupied households. In 1967, the numbers of owners and tenants were almost equal at 43.8 per cent and 43.7 per cent (Table 1.5 & Fig. 1.7).

Since 1978 the percentage of owners has grown for a combination of reasons, the main one being a policy environment that encouraged the expansion of ownership. In 1988, it was estimated that there were 11.2 million owners and 7.9 million tenants as a whole.

Table 1.4 The evolution of owner-occupation rates in various countries (%).

	Near 1950	Near 1960	Near 1970	Near 1980
Austria	36	38	41	52.9 (1983)
Belgium	38.9	49.7	54.8	62 (1981)
France	35.5	41.6	44.7	51 (1982)
West Germany		35.1	35.9	40 (1982)
The Netherlands	29.3		35.4	44 (1983)
Sweden		36.2	36.2	41 (1983)
Great Britain		41	49.2	60 (1983)
Italy	40	45	50.3	65 (1986)
Spain	49.5	50.6	64	69 (1982)

Source: INSEE Social Data, 1990.

Table 1.5 Distribution of owner-occupation in 1988 by age group (% of households).

Age groups	Owners	Recent owners	Tenants	Tenants w/o fee	Total
< 25	1.2	3.3	84.2	11.3	100
25–29	2.6	17.7	71.6	8.1	100
30–34	3.5	38.6	52	5.9	~ 100
35–39	6.7	48.4	38.7	6.2	100
40–44	11.4	49.4	33.7	5.5	100
45–49	20.3	43.6	30.4	5.7	100
50–54	31.1	33.4	29.8	5.7	100
55–59	45.6	23.7	25.8	4.9	100
60–64	56.8	14.8	23.2	5.2	100
65–69	61.9	9	23.7	5.4	100
70–74	54.1	3.4	32.6	9.9	100
75–79	58.9	1.6	27.3	12.2	100
> 80	47.9	0.8	35.2	16.1	100
All households	28.1	26.2	38.5	7.2	100

Source: INSEE 1990.

Levels and trends of land prices

When dealing with land prices, it is important to note the existing, and still growing, difference between the Ile-de-France, and in particular Paris, and the rest of France. French spatial differences will be discussed below. Paris experienced a typical real-estate boom in 1989–90, when there was an enormous gap between the figure of FF21,000 per m^2 for land suitable for construction (development land) in Paris intramuros (the administrative limits of Paris' 20 arrondissements) and the situation elsewhere. A recent study analyzed the way in which land prices had jumped and showed that in two years they tripled (Table 1.6). One consequence is that the growth rates of available building land in Paris exceeded those in the bordering departments. Nevertheless in 1990 land prices seem to have stabilized throughout the country and since then there has been evidence of a substantial retrenchment in values (EF 1990).

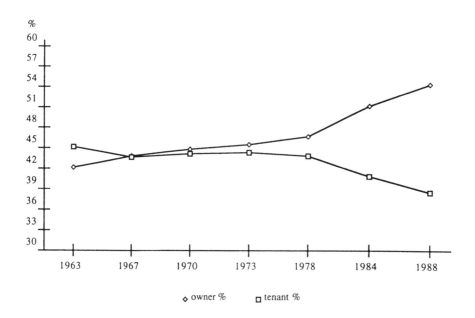

<table>
<tr><td>◇ owner %</td><td>□ tenant %</td></tr>
</table>

Figure 1.7 Housing tenure, 1963–88.

Table 1.6 Increases in the price of constructable land.

Departments of Ile-de-France	1986 (FF/m²)	1989 (FF/m²)	Rate of growth (%)
Paris (75th Dept)	6876	22587	228
Hautes-de-Seine (92)	1741	4457	156
Val de Marne (94)	770	1289	67
Seine St Denis (93)	609	781	28

Source: Etudes Foncières 48, 1990.

There is no homogeneous data in France that permits an analysis of land price patterns throughout the country. Work is currently under way through observations of the land and housing markets (see Ch. 5).

Requirements for land

One major 1980 study, known as the Saglio Report (Saglio 1980), calculated land requirements at that time. It brought to the fore again the land-use issue by examining the patterns of the production of land suitable for construction from 1975 to 1978. It encouraged a rational approach to space consumption. Furthermore, it estimated the need for developable land to be, on average, 50,000 ha per year. It also showed that the average collective habitat or multiple-occupancy residential development consumed less land (200 m^2 per dwelling) than the average one-family housing unit (2,150 m^2 per dwelling). In between, grouped habitat, mostly in the form of semi-detached housing, used on average 775 m^2 per dwelling.

Today, with the knowledge that different structures and trends apply to the Ile-de-France region compared with the rest of France, the figures, especially the average need of 50,000 ha, remain relevant. Some evidence of land requirements is provided in Figure 1.8.

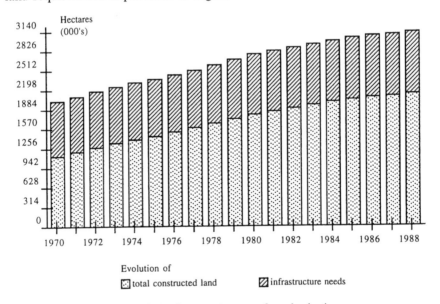

Evolution of
total constructed land infrastructure needs

Figure 1.8 Surface requirements for urbanization.

Figure 1.8 shows that land required for construction grew constantly from the 1970s, with a slight levelling out from 1986 onwards; this trend will probably continue. Another cause of land consumption in France is the creation of new towns. In retrospect this movement started long ago in the 1960s, when the innovative concept of new towns was imported from the UK. These schemes in France had strong political backing. In Ile-de-France (then called the Paris district), in response to problems of urban chaos and a perceived need to co-ordinate spatial and urban development, from 1961 to 1968 General De Gaulle delegated authority on urban matters to Paul Delouvrier. Ten new cities were proposed. In this respect the Paris region was an urban laboratory.

Although there have been some shortcomings (social and aesthetic issues) most new towns in the Ile-de-France region have been successful. With respect to demographic definition, most of them fall into the rural–urban outskirts category. Population growth in these areas has been notable. New cities such as Evry and Marne-la-Vallée are often top-ranked in terms of employment creation and economic dynamism.

Changing consumption patterns and major urban schemes such as the new towns have reshaped the spatial landscape of France. The result has been a substantial intensification of differences in economic and physical development between successful and unsuccessful areas.

1.5 Trends in spatial development

In order to identify existing differences in spatial development patterns at the national level, various criteria have been used, for example, population levels and growth rates. This has enabled a typology for the French territory to be presented.

National trends

Describing and explaining the trends of spatial development and the differences at the three levels of regional, urban agglomeration and city areas would require a lengthy study. Only the principal points and the latest findings in terms of spatial development can be summarized here.

Since 1947, a marked contrast between the dynamic Paris region and the rest of France has been observed (Gravier 1947). A rough outline of territorial boundaries shows how France is still divided between the "haves" and "have-nots" in terms of development and urban dynamism (Fig. 1.9). Using standardized criteria, two types of regions stand out. First, the dynamic ones in economic, social and cultural terms, indicated by the plus sign, are categorized by the numbers 1, 2, 3; the other less dynamic ones suffer from serious handicaps, indicated by the minus sign, and are categorized by a 4 or 5.

The first category is made up of the capital region Ile-de-France (1), the popular coastal regions (2) encompassing the Atlantic and the Mediterranean coastal areas, and the fluvial regions (3): the Rhône, the Rhine and the Loire hinterlands. Category (4), which has significant disadvantages at all levels, consists of an arc running from the Lorraine area to the Nord–Pas de

Calais. This industrial arc (producing coal, textiles, etc.) was formerly economically buoyant. Lastly, the arid diagonal region (5) belongs to the disadvantaged group; it is characterized by a lack of industrial development and urban dynamism.

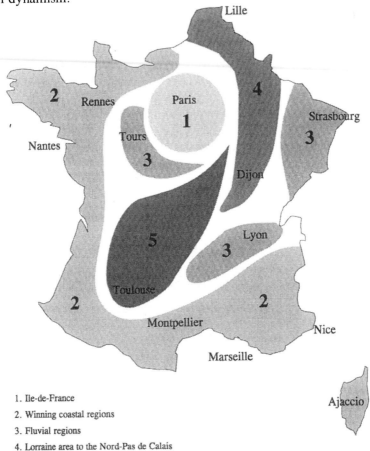

1. Ile-de-France
2. Winning coastal regions
3. Fluvial regions
4. Lorraine area to the Nord-Pas de Calais
5. Arid diagonal

Figure 1.9 Spatial development patterns.

Trends of agglomeration

Throughout the 1970s and at the beginning of the 1980s, most research concluded that there would be a massive process of suburban and peripheral growth. While in some respects this may have been true, the relationships have today become stable among agglomerations and between urban and rural areas.

In the 1990s, one large agglomeration, the Ile-de-France, predominates, with 10.4 million inhabitants. In addition, there are 12 agglomerations of more than 300,000 inhabitants that have European potential, but a vacuum exists between these two categories. In fact, only the Lyon agglomeration (and perhaps Strasbourg) may succeed in climbing the ladder of European competition. All 12 agglomerations have begun to exploit their strong assets, even though certain handicaps remain: in particular a lack of infrastructure and diffused and incoherent urban structures. Consequently, some side-effects have been noted, especially the length of time spent in individual and mass transportation (Merlin 1982). Taking Paris as an example, an average employee spends from 45 minutes to 1 hour on a journey from home to workplace.

As a result, vigorous pressures for suburbanization are now apparent. According to an INSEE study, although some dynamic small cities have had high growth rates, and the middle-size agglomerations are growing again after a period of stagnation, agglomerations of more than 200,000 inhabitants have acted as the "locomotives of urban growth". Once again, the Paris agglomeration heads the list with an average population growth rate of 0.5 per cent per year, leaving behind Lyon, Marseille and Lille. In absolute terms, strong population growth has been observed in agglomerations such as Toulouse, Bordeaux, Toulon and Orleans. In total, there are now 30 agglomerations of more than 200,000 inhabitants, compared with 28 in 1982.

Inner urban growth and peripheral urban development
In the 1970s and 1980s, most planning experts believed that rapid urban growth would take place in peripheral areas. Evidence from the census periods 1975–82 and 1982–9 showed that peripheral growth, taking into consideration demographic evolution, remained strong, with growth rates of 0.85 per cent and 0.89 per cent, respectively.

Table 1.7 Demographic evolution (000s of inhabitants).

| | Population | | | Evolution 1982–90 | | |
	1975	1982	1990	Total	Natural surplus	Balance
Continental France	52,656	54,335	56,614	2,279	1,820	459
Cities (perimeter)	23,565	23,413	23,540	127	862	(735)
Cities (periphery)	15,455	16,446	17,597	1,151	940	211
Suburban areas	7,827	8,746	9,687	941	147	794
Rural areas	5,809	5,731	5,791	60	(129)	189

Source: INSEE 1990.

The suburban population (known as *banlieusards*) was estimated at 17.6 million in 1990 (15.5 million in 1975). Demographically, *villes-centre* (city

CHAPTER 2
The policy environment

2.1 Goals and aims

This chapter presents the main guidelines of public policy with respect to land and property markets, as they are defined by the *Ministère de l'Equipement, des Transports et du Tourisme* (ministry of urban affairs). Policy-making with respect to land and property markets has undergone a complete change since 1982 as a result of decentralization. The growth of local powers, especially in the large cities, but also in departments and regions, has led to a much more intricate and intermingled set of public policies defined at each level. Given the growing differences in the spatial development of markets, it is important to make a distinction between public policy defined at central and at local levels, especially in the cases of major cities.

The main consequences of decentralization – the dualistic evolution of land and property markets, the increased flexibility of planning, and the exclusionary zoning inside and around major cities – has led the state to question the existing land and property development system. Its desire for change is stated explicitly through a series of recent laws and bills on the distribution of powers (Joxe–Marchand statute), property taxation and inclusionary zoning and linkage (*Loi d'orientation pour la ville*).

Seven topics characterize the policy environment in the 1980s:
○ institutional – the central–local relationship;
○ law – deregulation and land supply;
○ location of urban and industrial development and redevelopment;
○ the building industry;
○ economics – the 1987–9 property boom and its consequences;
○ the rate of property ownership; and
○ planning law – discretion versus matter of right.

The central–local relationship
The key event of the 1980s was the decentralization reform (1982), which followed two centuries of highly centralized power. The main responsibil-

27

ities in land-use and development have been granted to local authorities, especially the communes. This major policy has been led very firmly, with a general agreement of all political parties, and was not really questioned even after the 1986 parliamentary elections led to a change of government. Side-effects of decentralization have nevertheless begun to appear.

An important question raised at the end of the 1980s was the system of local finance. Currently this derives from a combination of grants from central government and local taxation, particularly land and property taxation. One of the main criticisms was that it did not allow enough cross-subsidization between poor municipalities, caught in the vicious circle of a low tax base, poverty, few economic activities, unemployment, etc., and the wealthier municipalities. The policy now implemented to mitigate these effects contains two important elements.

First, a recent statute has to a limited extent increased the cross-subsidization achieved through *dotation globale de fonctionnement* (grants from central government to communes). This is seen, however, as only a first step and it is recognized that the complete system of local taxation has to be changed.

Secondly, 1991 legislation (the Joxe–Marchand statute) provides some tools and incentives for improving the mechanisms for cooperation among neighbouring municipalities. Many agglomerations, particularly in the Paris region, are still split into a number of municipalities, causing serious problems in planning and development. Previously, *districts* or *communautés urbaines* (special cooperation boards) had been created on a voluntary basis by some agglomerations. The new legislation should encourage this.

Deregulation and land supply

A constant argument in supply-side land economics is that an over-regulated zoning system, by reducing the areas where development is authorized, is a major cause of land-price increases. According to this view deregulation and flexibility in planning should therefore improve the situation by lowering land prices. Important steps towards deregulation were adopted at the beginning of the 1980s, enhanced by decentralization, and extended in 1986–8. They did not, however, result in a stabilization or decline of land prices but rather in a dualistic evolution, with booming land prices in some areas such as the Riviera and the Paris region. At the same time, the side-effects of greater flexibility in planning were noticeable, especially the frequent modifications of local plans. Public policy is now changing and some measures are being prepared to improve the stability of local plans and increase the control of property development by the state.

The location of development

During the 1960s and 1970s, the government had to face rapid population growth and a shift from rural to urban areas. During those two decades, policy was mainly directed towards organizing and locating this growth, and planning policy was mostly defined in terms of urban containment and the development of new towns and of large industrial areas. The actions of DATAR (the government agency for economic and regional development) were designed to organize and redirect these developments, especially through economic incentives, to different parts of France.

The economic crisis of the mid-1970s and the changes in housing behaviour have altered this framework to a considerable extent. At present, the main features of public policy can be summarized briefly by stating that control by central government of the location of development exists but its efficiency is limited. Major firms enter into negotiations with lower-level public authorities, often major cities, and the state has no direct influence in the decision-making process.

A key public policy issue is government's response to the decline of certain industrial areas, characterized by derelict land and brown-field development land, e.g. the Lorraine region (see the Pompey case study, Ch. 6.2). In these areas, intervention of public bodies of different levels, including the state, is required before any real economic recovery is possible.

Where housing is concerned, the process of suburbanization, explicitly encouraged by the government though financial incentives in the 1970s, is now being questioned on various grounds, and a re-utilization of the existing housing stock is increasingly preferred to new development, which is costly in terms of infrastructure.

The building industry

After a rapid growth of construction cost from 1975 to 1986, the building industry benefits from low costs of material and labour. In fact, inflation (cost of living index) and construction costs have paced together. As in any other country, building is a major industry, employing 7–8 per cent of the active population (1.3 million). At a time when unemployment is high, and a major socioeconomic problem, it is obvious that building activity is encouraged by the government. At the same time, the incentives and subsidies for that purpose are a heavy burden on the state's budget. The total cost of housing subsidies is about FF120 billion, and there are some discussions and disputes between the ministries of finance and urban affairs, which have different objectives (see Part III).

Economics: the 1987–9 property boom and its consequences

In some areas of France, especially in the Paris region (Paris intramuros) and the western suburbs, the price of property (and land) increased rapidly from 1987 onwards. For instance, the price of land inside Paris trebled between 1986 and 1989, from FF6,876 per m^2 to FF22,587 per m^2, and the increase was not very different in the Department Hauts-de-Seine: from FF1,741 per m^2 to FF4,457 per m^2.

In terms of public policy, this price increase has had two major consequences: first, it has stimulated the intervention of a growing number of intermediaries in the land and property markets – *marchands de biens* (real-estate brokers, see Ch. 4.2) – who buy and sell options on property. Secondly the question of betterment or unearned increments in land and property values is once again to the fore. The land and property boom also resulted in increased difficulties for those seeking sites for *habitations à loyers modérés* (HLM – social housing), which is subject to the constraint of a ceiling on the land price. (For example the ceiling is around FF 800 per m^2 in the Paris suburbs.)

A higher price can be accepted if the municipalities or other public bodies agree to provide a subsidy, but the recent trends in land prices has made it more difficult or even financially impossible to build social housing in the western and southern suburbs of Paris. This exclusionary impact of rising land prices is increased by the attitude of some municipalities now responsible for granting building permits, which prefer gentrification to social housing and use whatever means are available to them (exclusionary zoning, use of pre-emption right, refusal of building permit on various grounds) to avoid the development of affordable housing.

The overall trend is thus the dispersal of social housing farther and farther away from central locations, and a growing disparity between the employment requirements of firms and the location of affordable housing. In response to this problem, 1991 legislation includes a set of measures, legal constraints and financial incentives designed to mitigate the phenomenon. The two main features of the statute oblige municipalities, under certain conditions, to include in their local plans a "local programme of housing" stating in detail the housing situation in the municipality and its needs in relation to the number of low-income households; and a financial mechanism for the cross-subsidization of housing. As with the system of "linkage" implemented in the USA (Alterman 1988), this mechanism creates the possibility of applying to new developments a special exaction, called *la diversité de l'habitat* (participation), through which a builder of non-subsidized housing has to provide up to 15% of the area for social housing or an equivalent contribution.

Property ownership

As stated previously, the rate of property ownership has increased in France and now stands at 54%. Up to the beginning of the 1980s, the growth of owner-occupancy had been regularly stated to be an objective of public policy. While this aim has not been explicitly abandoned today, a new emphasis has been introduced for at least two reasons. First, heavy subsidization of house acquisition at the end of the 1970s and the 1980s, after the 1977 reform of the housing finance systems, resulted at the same time in both a heavy burden for the state budget and a precarious situation for indebted low-income households, for they faced abruptly reduced housing subsidies as soon as their children reached the age of 18. A statute has been passed recently to solve the problem faced by "overindebted households".

Secondly, the mobility of the population is rather low, because reselling a dwelling during the acquisition phase, when a loan is still being repaid, is both costly and complex. A large rental housing market both reduces financial burdens and facilitates the geographical mobility of the population.

The government, facing a decrease of approximately 80,000 private rental dwellings per year, now aims to provide an incentive for household investment in property to rent. Special incentives are offered to low- and middle-income households that agree to rent for a long period (9 years minimum), at a "reasonable" rent (see Part III).

Land law: discretion versus matter of right

A last concern of public policy is related to land and property legislation and its use. The body of legislation has steadily increased since the Second World War. The set of laws, statutes and regulations that apply in different parts of the territory is intricate, fragmented and applied under the responsibility of different ministries, 22 regions, 99 departments and over 36,000 communes. Consequently its implementation is less and less satisfactory. It is no longer possible for any of these bodies to understand or even to be fully informed of the present state of legislation. The *Conseil d'Etat* (supreme court) has recently been asked to undertake the special task of "cleaning" the set of regulations, but the logic of "autocomplexification" of written law is powerful, and is now increased by the emergence of new European Community legislation.

The main feature of the urban property market is the increased competition between regions, departments and cities, both to attract economic activities, especially of a high-tech character, for employment and resource reasons, and to improve urban housing and environmental conditions in order to give a favourable image to the city.

The regions, departments and municipalities do not intervene in the housing sector except through social housing companies (HLM); public intervention in favour of housing comes only from the state. The national system is characterized by four kinds of subsidies: first, subsidies to build or to

31

renew rented social housing and to develop ownership by low-income households; secondly, allowances granted to households according to their incomes and housing expenditures (both rents and loan repayments); thirdly, tax deductions to households that want to become homeowners or want to invest in the private rental sector; fourthly, subsidies and tax deductions to encourage saving in particular financial assets that are then used to finance housing loans at a low interest rate.

On the whole, the system promotes the new housing market. Since 1984 successive governments have emphasized the social component of their housing policies. Subsidized loans for ownership are increasingly reserved for underprivileged households. The income ceiling for entering HLM housing has not been re-evaluated and so these dwellings are now reserved for very low income households. In addition, the high-income households cannot benefit from tax deductions when they buy their own homes.

A goal of government policy is to reduce budgetary expenditure in the housing sector and consequently the increased social content of their policies is one of the means used to reach that goal.

Another policy tool is the deregulation of the housing markets, of the housing loan market and of rent control. Deregulation has been achieved for rents, except in Paris where strict control still applies. The consequences of this deregulation are ambiguous. On the one hand, by raising the profitability of renting one's property, the attractiveness of renting property is enhanced. The logical consequence of this is not yet apparent since the number of privately rented dwellings is decreasing by 80,000 a year. On the other hand, deregulation, as previously explained, has caused increased segregation and its correction is clearly a priority for the present government.

PART II
The urban land market

CHAPTER 3

The framework and functioning
of the urban land market

3.1 The legal environment

The trend towards centralization to the late 1970s

No comprehensive legal frameworks for urban land-use were instituted in France until the 19th century. The right of pre-emption goes back to the Ancien Régime: through it a feudal lord was allowed a certain period for buying real estate on the market before a deal was concluded between two parties. This institution enabled feudal lords to select the new vassals to reorganize the land-use of a given domain and to control local taxes: it was known as *retrait féodal* (feudal restriction). Although the Revolution suppressed this right, it appeared later on under a new label (fiscal pre-emption right) as a way of generating funds for the state. Later 19th century attempts to establish pre-emption procedures contributed to the achievements of Haussmann.

Urban theory gave further impetus to urban legislation at the turn of the 20th century (Tribillon 1991: 38). The principles of hygienists, urban planners and sociologists were progressively implemented. The Urban Act of 1919 entitled local authorities to prepare local plans. However, only 273 of 1,900 towns had prepared a local plan by 1939. During the turbulent years of the Second World War, the Vichy government took hold of the urban situation and its processes by issuing a decree (Decree of the 15 June 1943) that aimed to control, conceive and implement urban planning at all levels of government. This meant that the state administration centralized demand and regulated supply, especially of the conversion of land to first urban use.

At the end of the Second World War, new institutions confirmed this process and continued extending the powers of the state. In particular, through a series of laws and decrees in the period 1953–4, the state explicitly acquired the power to expropriate land (particularly for carrying out reconstruction projects) and to specify the professionals (architects, economists, civil engineers, etc.), who could be in charge of such projects. Consequently, the public sector predominated through comprehensive projects known as *zones opérationnelles* for housing (ZH) and industrial (ZI) development (Tribillon 1991: 54). Urban models such as the "Grille Dupont" were supposed to foresee future demand so that the required infrastructure could be built in advance.

Because of reconstruction efforts and urban growth, the government conceived and put into effect the first comprehensive urban policy in 1958, at the same time consolidating its legal framework. Fourteen ordinances, covering planning and land-use control, were part of the legal core. Since certain gaps remained, for example in respect of public land-banking, additional legislation in 1962 created another land-policy tool, known as the "right of first refusal" or pre-emption right. Thus urban jargon was generated to accompany such procedures as ZUP (priority zones for urbanization) and ZAD (deferred development zones).

In brief, in the postwar years the state adopted zoning in order to control land-use and to foster urban development. These measures broadly reflected a traditional preference for regulatory policy-making rather than taxation strategies in finding a "correct" land-use. For instance, one of the first tools, the ZUP, established in 1958, was established by municipalities in accordance with government policy, which designated land to be urbanized. Under ZUP procedure mayors were granted pre-emption rights for a period of four years. The procedure was intended to complete and accompany the expropriation right. Public bodies (cities, SEM, etc.) concentrated all their efforts in infrastructure matters in the large areas that were urbanized.

Other significant laws followed. On 31 December 1967, for example, parliament approved the *Loi d'orientation foncière* (town-planning law). Now considered to be the cornerstone of urban legislation, it established a two-tier planning system: the *Schéma directeur d'aménagement urbain* (SDAU or SD: master plan) and the *Plan d'occupation du sol* (POS: local plan).

Master plans for individual agglomerations were designed to make forward projections of their evolution, analyzing employment patterns, demography, infrastructure, etc., and to plan for and provide guidance for the development of the facilities and *équipement* (infrastructure) requirements of all public and private entities. The master plan was specifically conceived as the key or basis of the new legal framework. Accordingly, the local plan was only meant to be a micro-level tool for the implementation of the aims and objectives of the master plan. (Fig. 3.1).

To strengthen this planning system, another reforming land-policy law (31 December 1975) provided local government with pre-emption prerogatives and a set of legal tools. In addition, decentralization brought about a redistribution of land-use planning and urban development control rôles during the 1980s. The laws of 7 January 1983 and 18 July 1985, and application

35

decrees of 1986, refer to the sharing of powers between the municipalities, departments, regions and the state. The decentralization law introduced new principles and notions concerning urban development and public land-banking (see below). By 1990, 193 master plans had been approved out of 413 drafted. Over 5,000 municipalities were involved, covering a total population of 23,175,000 (Comby & Renard 1990: 168). Over 13,000 local plans, out of a possible 36,500, had also been approved by 1990.

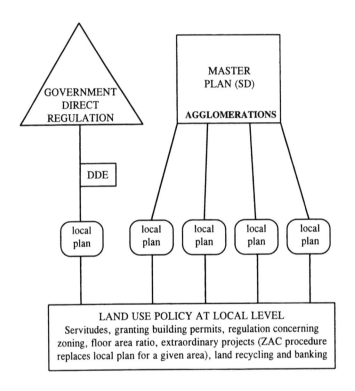

Figure 3.1 Master-plan implementation.

Decentralization effects and the present hierarchy

Decentralization has brought about significant changes in land-use planning at all levels of government (Wilson 1988). On 1 October 1983 the power to control land-use through the local plan (POS) was conferred on municipalities. This local plan procedure was intended to incorporate all ordinary urban planning and other day-to-day matters in a modest manner (Tribillon 1991:

46). Because of the partial failure of master plans, local plans have become predominant since they contain clear information on projected changes in urban areas.

Some incoherence and divergent practices have surfaced among municipalities that have wanted to exercise their right to control their own urban space, creating various chaotic situations. Policy-makers envisaged intervention by higher tiers of administration, especially when municipalities compete for industrial development and aim to attract innovative firms. Nevertheless, only persuasion and a common interest will lead to urban co-operation among municipalities. The state's rôle is to safeguard the national interest. At present, both legislative bodies are discussing setting up a more balanced policy, based on the Joxe–Marchand statute, which is designed to reduce urban incoherence through intercommunal co-operation in land-use and urban planning.

Likewise, the 1991 *Loi d'orientation pour la ville* (law on urban development) provides for state intervention if municipal action is non-existent or there is a persistent doubt about a municipality carrying out its responsibilities. Broadly speaking, it seeks to increase social housing and to create a basis for fiscal cross-subsidization among municipalities (see Ch. 3.2 and 3.3).

Environmental laws and protection regulations

The environment (parks, forests, reservoirs, historic sites, etc.) has been at the centre of attention in the ministries of agriculture, environment and urban affairs as well as in parliament. Forests cover 24 per cent of France, amounting to over 14 million ha (10 million ha in private ownership, 1.7 million ha state-owned and 2.5 million ha owned by local and other public authorities). A national forest agency (ONF) is in charge of management, erosion prevention, conservation of springs, rivers and landscape protection, and the reforestation of wooded areas. If an owner wants to clear a piece of his wooded land of over 4 ha, he must ask the permission of the ministry. Since 1960, seven national parks have been created. Designation procedure includes public inquiries and consultation with local authorities, but in practice designation depends on the decision of the *Conseil d'Etat*.

A law dating back to 1913, extended and improved in 1930, regulates historic and other picturesque sites. The aim is to protect and restore historic buildings and other valuable sites. In 1976 a law (Nº 76-629) was passed and with a decree instituted the notion of "natural reserves" in order to protect natural sites that include "special" fauna and flora. The environment ministry decides which sites are to be considered "natural reserves". In the event of opposition by land-owners, the supreme administrative court makes a final decision. Once a reserve is designated, further development is forbidden and

37

other recreational activities are restricted.

Other laws protect and regulate the utilization of mountain sites (L-9-01-1985), monitor (and if possible, reduce) construction in coastal areas such as dunes and beaches (L-3-01-1986). Other laws are intended to curtail the massive expansion of billboards (L-21-12-1979, the 1980 decree) in the countryside, permitting only travel and public assistance announcements.

Authorities and lawmakers have toughened their stand against water and air polluters. The fight against water pollution has been advanced in two phases by the creation of a National Water Committee in 1964 and the rational utilization of the six water basins. A 1966 decree introduced an innovative financial scheme in order to raise special taxes to finance investment and operation costs. The "polluter pays" principle was also implemented and as a consequence monitoring of industries through impact studies began.

In the 1960s, the air pollution law (L-2-08-1961) was passed, followed by a specification of technical standards. A technical observatory, the Air Quality Agency, within the Ministry of Environment, is charged with monitoring air quality.

Nuisances and noxious elements have been regulated since 1917. A strict procedure has been followed by prefects, who have lists of dangerous activities that are not permitted because of their effect on people and the environment. In the 1970s, this procedure evolved. Authorities carried out studies in order to analyze the effects of development on the environment, landscape, fauna, etc. Measures are taken so that negative effects are minimized and if possible eliminated. In general, policy-makers have gradually sought to bring legislation into line with the standards required under EC Directive EEC (85) 337.

Local plans: the implementation of a POS

Municipalities have taken advantage of their increased powers over land-use. After 1 April 1985, where local plans exist, municipalities were empowered to deliver building permits and other related land-use authorizations, for demolition, refurbishing, etc. By 1991, 13,300 municipalities out of 36,000 had published local plans (i.e in operation). Of these, 47 per cent are municipalities with more than 20,000 inhabitants; all communes with more than 50,000 inhabitants have now published a local plan. Increasingly municipalities have rushed to prescribe and implement local plans.

Figure 3.2 presents the different steps and the various actors in the preparation and implementation process of a local plan. Today, a local plan (POS) has three functions:

○ to record developed urban zones
○ to identify future urbanized zones (NA)
○ to limit development on non-urban zones (NC, NB, forests, etc.).

In practice, the local plan must also indicate the socio-economic context of the municipality and present an interpretation of each individual plot and its boundaries (e.g. FAR, housing zones, economic zones, wooded areas).

In a local plan, a distinct typology exists: urban zones (marked with the initial U), zones to be developed (NA) and zones with a natural character (the rest). Building permits may be obtained without major difficulties in urban areas (U). Land for future urbanization is called NA (*non aménagée*). Areas where construction is restricted are divided into two types: areas where construction is forbidden (NC areas, mostly agricultural land, and ND or conservation areas); and NB areas, where construction is allowed as long as building is low-density (usually in rural areas).

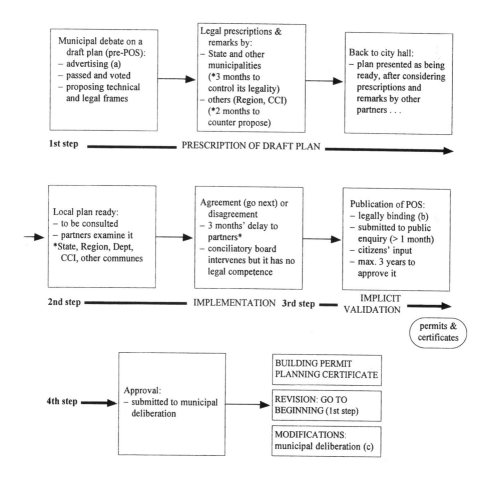

Figure 3.2 Planning process for local plans.

39

A local plan's implementation involves four legal steps: prescription of draft, preparation, publication and approval.

The municipal council first discusses and approves a draft plan (with its technical and legal mechanisms, involving different organizations). The project is then transferred to the prefect, who checks its legality and its conformity to national interests (defence and public projects). The prefect may define the state intervention policy and inform the mayor of the different competent services the state could offer (DDE among others). The prefect's scrutiny takes three months. Meanwhile, the mayor sends the draft plan to other public or private organizations (e.g. CCI, region, neighbouring communes) which have up to two months to respond or comment. If a conflict arises, the administrative courts can intervene if asked to do so by any of the above actors.

Secondly, once the local plan is technically ready, the state and other partners make observations and remarks. If all partners and interested actors have agreed on the main principles after a series of studies of impact, the mayor can proceed to the next step. If any of the actors (especially neighbouring communes) or partners oppose the plan, there is a three-month period within which to make counter-proposals. If it is impossible to establish a middle ground, a conciliation board may intervene. But this body has no legal jurisdiction and it may be obliged to refer the issue to the proper authorities (administrative or civil courts, see Ch. 1.1). Although referral to the courts in such cases is still rare, the phenomenon is bound to grow. In fact, more local plans are being stopped for reasons of substance rather than procedure.

Thirdly, if no conflict has arisen, the plan is published. This step is very important since it implicitly validates the process of granting building permits. In essence, publication enables the mayor to grant building permits and other legal documents for subdivisions, etc. The local plan allows the planning department to control urban activities and puts land-use regulation into effect. In effect the local plan operates as if it had been validated. The mayor has three years in which to have the plan approved by a majority vote of the municipal council; in the meantime the de facto situation is in practice almost irreversible since it would be very costly to change it, and the financial and political stakes are substantial. Until publication, public input is minimal since public inquiry officials have insufficient powers and their propositions can have little or no effect.

Finally, the municipal council approves the local plan at any point in the duration period (three years). Subsequently, an approved local plan can be modified by only two procedures: revision and modification. Revision follows the same steps as above. Modification, by contrast, according to the procedure introduced by the 1983 law, allows for changes of the local plan by a simple deliberation at the city hall after a public inquiry. The general principles of the local plan must however be respected. In particular, wooded areas should not be reduced. In brief, the notion of flexibility of local plans is becoming more widely accepted.

The law on land transactions and first urban use affecting the private sector
Private law relating to land transactions has a source different from urban planning law, and different usages apply. This section discusses the processes of selling and buying. There are two distinct steps: engagement and commitment (or affidavits), and the signature before the public notary that legally confirms the transfer.

As a first step, the prospective buyer and seller of a property agree in a document known as an engagement/commitment affidavit a fixed price and the document's validity. This can be regarded as the "real-estate contract". If there is a bilateral agreement, neither party may retract this legal commitment except on grounds of illegality or public interest. The agreement has a set duration (two weeks to three months) and is backed by a financial deposit. The sale can then be completed, with the signing of the mortgage deed, according to Articles 1583 and 1589 of the Civil Code (which requires perfect agreement on the "object/item" and the "price"). Very often, when a real-estate agent intervenes as a third-party representative, this step is known as the *compromis de vente* (sale engagement).

The initial commitment act must be signed in due form by both parties or their legal representatives. In addition to the time limit to be applied, it should state certain conditions that will render the commitment void (e.g. failure to acquire an urban certificate, building permit or financial support). It also designates the appropriate public notary for signature of the authentic mortgage act (*hypothèques*). The vendor cannot go back on the agreement to sell during the set period of time and, if the purchaser fails to close the purchase during this period, he will forfeit the deposit (usually 10 per cent of the total value). If everything has been cleared, both parties go to the assigned public notary in order to register the transaction.

The *acte authentique* (mortgage deed) forms the second step. Once signed, under the supervision of a public notary, the deed will be sent to the mortgage office. It contains valuable information about the owner's civil status and age, and details of the property. Notarial fees and other related charges are calculated. In addition, formal advertising of the mortgage (*publicité foncière*) is paid for. At this point, all other fiscal charges must be met (see Ch. 3.3).

Other types of transactions (e.g. first urban use such as subdivision of property) follow the same legal steps. It is necessary, however, to point out that a developer, selling lots to individuals, can do so by delivering dwellings that are ready for occupation, or by selling for future delivery (with timetables for payments and construction laid down) or for long-term sale. In the latter case the developer signs a contract to deliver a dwelling for which he will get paid on completion of construction. In all cases, the eventual buyer will deposit a given sum, usually 2–5 per cent of the total selling price. Certain guarantees must be met. A preliminary contract or contract to reserve the property is binding on buyer and seller, and serves the same purpose as the sale engagement and commitment deed described above. The reservation contract should state the general parameters (area, boundaries, lot numbers, servitudes etc.), the sale conditions and its eventual conclusion (second step). The sale contract, notarized by a public notary, is as described above.

Co-ownership and private *servitudes* (easements or covenants) may apply. Among private parties, a valid standard signed document is available, stating conditions applicable to such things as shared walls and drainage. Co-ownership agreements between apartment owners are often made: they lay down conditions for use of common parts (such as lifts) and their maintenance and replacement. Sometimes servitudes can apply to subdivisions, for example, for public space managed by co-owners or local citizens. Given that the commitment deed (either to buy or to sell) is an important element of the acquisition procedure, real-estate brokers have intervened to create "options" on land and property purchase that can be traded (see the discussion of *marchands de biens* in Ch. 4.2). These are unilateral *compromis d'achats* (commitment contracts) for a fixed period, especially for non-building (NC) land, using the condition of substitution (*faculté de substitution*). Since the broker is not the final buyer, the option can be traded on speculatively. This speculative market depends on the belief that local authorities will allow specific building projects to be negotiated. This involves the reclassification of land from NC to NA zoning, in which case the public mortgage advertising act is not invoked by the authorities. The broker aims to negotiate land reclassification from NC to NA with the public authorities so that urbanization can follow. If the broker succeeds, the commitment contract is resold to a developer who will take up the option. Often, this chain can be extended for a couple of months or even years.

This process has led to the creation of parallel markets in some agricultural areas, with speculative (or hope-value) prices above normal agricultural land price levels producing a "ratchet effect" (see Ch. 5.4).

Instruments for plan implementation

Plan implementation is obviously the key to an assessment of the overall efficiency of a planning system and the avoidance of unwanted side-effects. In France, implementation is accomplished via a set of legal and financial devices. The main instruments concerned with the urban land market are examined here, showing the process from zoning to development, building permits, development freeze, expropriation, pre-emption, improvements and their charges, betterment levies etc. Private law relating to land transactions has already been referred to. Before going into details, the principle of private property must again be highlighted.

The civil code (Article 544) states that ownership is "the right to the use and abuse of property in an absolute way, except if it is in contradiction with the law and regulations". Article 545 notes that "nobody can be obliged to give up his property, except for reason of public interest, and then only under the condition of fair compensation." Except in case of expropriation (see below), property may be exchanged and sold freely. The authenticity of a transaction is guaranteed by the involvement of a public notary.

From zoning to effective development There is no systematic link between a local plan and its effective implementation. The designation of an NA zone, designed for future development, has no direct effect on the land-owner except that such a designation usually increases land prices. There is no systematic replotting device, nor any process by which the local authority can oblige the land-owner to develop and build, (unlike, for example, the situation in Germany). This also applies to unbuilt U areas.

In retrospect, it is clear that policy-makers regarded public land-banking as a way to ensure effective development, but land-banking has rarely been used systematically. Land-banking usually refers to the procedure whereby land is bought and stocked for a future use that is not necessarily determined at the time of purchase, either by the public or private sector. The first wave of public land-banking occurred in the 1950s through the creation of a special fund, the *Fonds National d'Aménagement Foncier Urbain* (national fund for urban development). This granted aid and heavily subsidized loans to the government itself and to local authorities to buy land reserves. It was enhanced in 1962 by the creation of the pre-emption right applicable in "deferred development areas"(ZAD).

In 1975, another legal tool was added: *zones d'intervention foncière* (ZIF – land-banking zones). This procedure had two aims: to recoup betterment, and to provide an appropriate management tool for urban areas. Its pragmatic approach has resulted in the creation of green spaces, the construction of affordable housing, the provision of urban services, and the renovation and restoration of old urban environments.

Throughout the 1970s the government set up several public land-banking agencies to implement large-scale land-banking schemes. For example, the *Agence Foncière et Technique de la Région Parisienne* (AFTRP – Paris Region land agency), created in 1962, has bought many thousands of hectares in the Paris Region, notably through pre-emption, and has played a key rôle in the preparation of the new towns there. It has had a great success in keeping land prices at a low level in spite of severe pressure. (See Ch. 10.2 for further details of the ZAC Citroën case study and AFTRP's rôle as a land supplier).

Other examples include two *établissements publics* (public land development agencies) that have been created in the Languedoc–Roussillon and Aquitaine Region to prevent speculation in tourism development in coastal areas. Two other public land banks have appeared: in 1967 in the Normandy Region (*Etablissement public de la Basse-Seine* – EPBS), and in 1973 in the Lorraine Region (*Etablissement public de la métropole Lorraine* – EPML). They are financed by the *taxe spéciale d'équipement* (TSE), a special tax levied by the local authorities, who also have majority representation on the boards of the agencies (see the Lorraine case study, which refers extensively to the EPML's policy, Ch. 6.2).

Local authorities have also received some funding for the implementation of land banking within their areas. In practice, cities have not had very active land policies, except in a limited number of cases, despite the tool of the urban pre-emption right. Among the exceptions, the city of Rennes in the west of the country has had an extensive "Swedish" anticipated land-acquisition programme (see the Rennes case study, Ch. 6.1).

The progressive exhaustion of public land reserves in the 1980s and the recent boom in some land and property markets have once again encouraged the idea that land-banking can be profitable (in the broad sense); recent legislation allows for the creation of land banks *établissements publics fonciers* (land banks) by major municipalities.

Instruments to protect planning Two legal instruments are intended to protect planning: a mechanism for deferring decisions (*sursis à statuer*) during the preparation of a plan and the delineation of *emplacements réservés* (frozen areas) in a legally binding local plan. When a plan is being prepared, a building permit application can have a deferral applied to it for up to two years, if the building could "compromise or make more costly the implementation of the plan". Abuse of this tool has occurred and courts are increasingly anxious to check the validity of such a decision. It is also possible to use such deferral in other cases, for example, as soon as a public inquiry preceding an expropriation has begun, or when a public project has been drafted by a public authority (see Ch. 4.3).

Moreover, a local plan usually includes frozen areas, which are normally those earmarked for future construction of public services such as schools, roads, hospitals and public parks. In such areas, no building permit can be granted to a land-owner. The land-owner can, however, ask the municipality to buy the land. The municipality must respond to such a request within one year. If the municipality refuses to buy, the area is then "unfrozen" and the local plan has to be changed accordingly. Alternatively, if the municipality buys, the price is fixed by the court in the same way as with the expropriation procedure. The owner can appeal to the courts to obtain a fairer (higher) price, but this is a fairly lengthy process.

Expropriation The expropriation mechanism is the legal process whereby a public authority is entitled to deprive an owner, after payment of due compensation, of property in order to achieve a public objective. This process is divided into two phases. First, an administrative phase ascertains whether the proposal is "in the public interest" in the first place. A *commissaire-en-quêteur* (commissioner) conducts a public inquiry and reports his conclusions; the resulting *déclaration d'utilité publique* (DUP – notice of public interest) is then taken over by the state administration. Often, this means the prefect, but for major projects such as land acquisition for *trains de grande vitesse* (TGV) routes the Conseil d'Etat may be involved.

The second judicial phase focuses on the assessment of property values. The compensation to be paid is the full market value, plus a re-use allowance that is often as high as 25 per cent of the market value. When the property owner disagrees with the price being offered by the expropriating body, he/she may ask the courts to review the price. Despite increases in prices usually granted by the courts, a large majority of land-owners accept the proposed price because of the long delay, often two to three years, required to get a judgment. During the 1950s and 1960s, the scope of expropriation was greatly enlarged to include housing development, and the notion of public interest was increasingly assessed by cost–benefit analysis instead of being decided according to the nature of the project (road, airport, etc.).

Available data suggest that about 120 km^2 per year are submitted to the compulsory purchase procedure (Table 3.1). The trend is now to reduce the amount of expropriation. Most current cases of expropriation are for "classical" purposes: urban public services, nuclear plants, highways or TGV tracks. Agricultural and natural spaces are noticeably more affected by the expropriation procedure than urban land (see Table 3.2).

Local authorities remain the principal beneficiaries of this procedure. For major projects, the cost of land is only a minimal percentage of the total cost and the expropriating body may therefore be tempted to propose prices well above the prevailing market price in order to avoid opposition from land-owners. But this overpricing can have some damaging consequences on the agricultural land market (see Ch. 4.1).

Pre-emption The pre-emption right – the right of an authority to claim priority to buy a property when the owner has declared his intention to sell – has been widely developed in the past 30 years and is probably a feature particular to French public land policy. Any decision to exercise the right of

45

Table 3.1 Land expropriation in France.

Year	25 départements (results) surveyed		France (estimate)	
	Number	ha	Number	ha
1983	810	3,168	2,765	8,111
1984	694	1,1901	2,547	18,330[a]
1985	762	2,773	2,861	11,240
1986	677	2,819	2,805	9,905
1987	626	3,469	2,431	12,833
1988	671	3,444	2,566	12,222

Source: Etudes Foncières 45.
Notes: a. The figure for 1984 is distorted by the declaration in the département of Orme of 10,026 ha of DUP. A more sensible figure for France would be 8,304.

Table 3.2 Land-use category in the expropriation process (%).

Year	Agricultural land	Urban land
1983	85	15
1984	91	9
1985	84	16
1986	84	16
1987	83	17
1988	89	11

Source: Etudes Foncières 45.

pre-emption must be explicitly motivated by a project defined by law as in the public interest (Urban Code, Article L-300-1).

There are four main pre-emption rights and their implementation is subject to technically complicated regulations. The first land pre-emption right was introduced at the beginning of the 1960s within "sensitive natural areas" in order to prevent speculation on coastal areas, such as the Riviera. It is now applied in most coastal areas. Managed by the departments, it is financed by a special additional to the land-development tax (see Ch. 3.3).

Secondly, in 1962 the state created the broader pre-emption tool: *zones d'aménagement différé* (ZAD – deferred development areas). It created a time limit and defined the affected zones. The time limit was initially eight years, later lengthened to 14. Within this time period, the pre-emption right was open to a public authority (state or municipalities), at the market price prevailing one year before the creation of the ZAD. The price is, however, subject to the land-owner's seeking to have it revised by the court. This major tool has been widely used, especially in the Paris region, in peripheral areas subjected to strong urban pressure.

It is erroneous to suggest that ZADs have frozen land prices, but it is obvious that they have smoothed their fluctuation, limited speculation and facilitated development control. They have been extensively used in new towns. Under this procedure, urban development is unlikely to occur for several years, so the policy allows for deferral of land acquisition. Today, the ZAD procedure still applies, but its use is restricted to municipalities without a local plan (POS).

Third, the *zone d'intervention foncière* (ZIF – urban pre-emption right) was introduced for urban areas in 1975. In contrast to the ZAD, the purchase price by the public authority (usually the municipality) was supposed to be the market price.

Fourth, in the 1980s, in order to simplify the legislation, an innovative procedure called the *droit de préemption urbain* (DPU – urban pre-emption right) replaced and extended the aims and scope of the ZIF for pre-emption. The DPU enables a municipality to acquire land or buildings within a given perimeter. This right can apply to all urban zones (classed U in the local plan) as well as prospective urbanized areas (classed NA) once the local plan is published. If disagreement occurs, the expropriation procedures can be enforced. But if the municipality does not exercise its pre-emption rights, the particular property is free of DPU procedures for a period of five years. The DPU is extensively used by large municipalities, including Paris, and also by medium-size and small municipalities. Most of the case studies (see Ch. 6 and Ch. 10) relate to areas subject to this pre-emption right.

In an area subject to the pre-emption right, a property owner who intends to sell a piece of land, a flat or a house must announce his or her intention to the municipality, stating the asking price. If the municipality intends to buy the property, it must make this known within two months. If it can accept the proposed price, then the transaction takes place. Alternatively it can make a (lower) counter-proposal. The owner can then accept it, abandon the sale, or defer the case to the courts in order to seek a fairer price. For built property, the price to be paid is the full market value. For land, it is the present price in the use it had at the last revision of the relevant POS.

In 1985, the urban pre-emption right was revised. In 1986, the movement towards deregulation resulted in a relaxation of certain pre-emption rules, but a 1989 law has adopted a more pragmatic position, notably on rural land. The state is expected to take the initiative to implement pre-emption procedures on zones classified NC, where strong price pressures are apparent. Parliament has however finally decided to let each municipality deliberate whether the DPU can be implemented. Today, a large and growing number of local bodies have adopted the DPU procedure.

On the whole, the DPU is useful in improving the efficiency of land policy. It provides the municipality with some fresh information on the state of the market and it helps to curb speculation in periods of boom. Most importantly, it is possible to buy strategic properties at a "reasonable" price as soon as they come onto the market. Controversy nevertheless arises because of the large discretionary powers granted to municipalities which can sometimes be used for hidden objectives. For instance the municipality may exert pressure on the seller in order that he or she sells to a builder chosen by the municipality. Another criticism stems from consequential delays in the transaction, especially when the case is deferred to a court, because the fluidity of the market is reduced.

Improvements and their charges (TLE, PAE)   Urban development requires the provision of certain facilities (roads, sewerage, parks, schools, etc.) financed by a public authority (i.e. the taxpayer). Alternatively, such facilities can be financed by the developer/builder (i.e. in effect the household that buys a new dwelling, or the land-owner who benefits from the increase in land value resulting from the construction of such facilities). After two decades of wavering, policy-makers have tried two methods of regularizing the position: a systematic fiscal device and case-by-case arrangements. In the first case, the *taxe locale d'équipement* (TLE – local development tax)) is calculated as a percentage (usually 1 per cent) of a predetermined lump-sum value of the floor-area to be built. Case-by-case agreements are negotiated between a developer and a municipality. This is more common in comprehensive development areas (see discussion of ZACs, above).

Besides the ZAC framework, new tools called *programmes d'aménagement d'ensemble* (PAE – special exaction areas) were introduced in 1985. The PAE mechanism is situated halfway between the two methods referred to above. The rapid growth of this mechanism enables us to understand its effects on land markets. Whenever the PAE is to be implemented, the municipal council delineates the area within which developers who have secured a building permit will have to pay all or part of the costs of the public infrastructure required by present and future residents. The amount of the exaction is not preset: the law states only that the amount cannot exceed 100 per cent of the real cost of the infrastructure. At the outset, the municipality must establish the nature of these public facilities, their cost, the time schedule for building, as well as the developers' share (per m^2 of floor area to be built) of the total cost, and how the cost will be apportioned among the various types of buildings. The exaction can be in kind or monetary.

If the local authority does not implement the PAE infrastructure programme on schedule, the developer can demand to be reimbursed for exactions already paid. This legal framework is thus flexible and consistent with the principles of decentralization. For municipalities, the main constraint is that the time schedule for providing facilities must be decided at the time when the special exaction area is approved and this is independent of the real timing of the development.

This tool is too recent to be evaluated thoroughly. Compared with the former practice of informal, and often illegal, case-by-case exaction, this new procedure has clearly improved the system by clarifying the statutory basis of exactions. Moreover, the announcement of the PAE programme, displaying the real servicing costs, could have some influence on the land-supply curve, and act as an incentive for land-owners to lower their supply prices in given demand-curve situations. This is, of course, an intuition that results from a set of convergent observations.

Taken as a whole, the special exaction areas mechanism seems to be an interesting adaptation of public–private partnership providing at the same time some flexibility in the management of the development process and a forecasting device that allows developers and builders to plan their investments. The aim of the PAE method is to render the operation of the urban land market more transparent. The products of these various contributions for improvements are discussed in Ch. 3.3.

Betterment levy: the legal density ceiling Introduced in 1976, the betterment levy mechanism is known as the *plafond légal de densité* (PLD – legal density ceiling). It recoups all or part of the increase in land value resulting from the development rights attached to a piece of land. The PLD sets a limit (the ceiling) to land-owners' development rights, independently of other constraints resulting from regulations. The land-owner, when he gets a building permit with a floor-area ratio greater than the stated ceiling, must "buy" from the authority the development rights that are in excess of the ceiling and must pay a fee equivalent to the market price of the extra land that would notionally be needed in order to avoid exceeding the PLD.

The PLD aims to reduce the increases in land values in centrally located plots with high densities and at the same time discourage high-density schemes by fiscal means while bringing extra resources to local councils. Thus it provides a mechanism to recoup windfall gains resulting from zoning. In 1983 and 1986 two statutes softened the PLD's effectiveness and its rate is now decided by each municipality, the sole constraint being that it must be greater than one. It is not therefore used as commonly as it was before, but is still an important source of income for local councils, especially in areas where large office development occurs, for example in the western municipalities of the Ile-de-France. Around 2,000 municipalities throughout the country have kept the PLD mechanism in operation. As a result, betterment levies have reduced land prices. In central Paris, the PLD drove prices down by 30 per cent in the year it was introduced. The city of Paris cancelled the PLD in 1987. Elsewhere it is now more or less a supplementary tax on large-scale development on top locations. It is not clear what effect it has on the land market.

Information systems

The system of information for land and property encompasses many different elements that may be classified according to various dimensions: public–private, individual versus statistical, legally binding as opposed to informal, objective versus subjective assessment, etc. The available public information (cadastre and property registry) is described briefly below.

The cadastre The cadastre, as in many other countries, provides information about ownership, boundaries and fiscal value (i.e. administratively assessed rental value). Instituted as early as 1807 in order to establish property taxation, the cadastre was completed in the 1850s. It is still the responsibility of the finance ministry. It contains two kinds of documents: maps and the property register. The cadastral survey contains maps on scales as large as 1:500 permitting accurate identification of any plot of land, each of which is numbered. Both cadastral maps and the property register are available to the public in town halls, as well as in the finance ministry. The cadastral system makes it possible to find the location of any plot if its code number is known, or to find a plot in the file, provided that its location is known, and to obtain a list of all the plots owned by a given land-owner.

The cadastre is now computerized, but its management is costly in relation to tax receipts (see Ch. 3.3). It is considered to be quite reliable, but updating after a transaction is still a rather lengthy process: it often takes about a year for a change of ownership to be recorded.

The fichier immobilier (property register) or mortgage record office The property registry is maintained by a branch of the finance ministry and it includes information about every property transaction, with copies of deeds, information about the nature of rights to property, the dates of deeds, and the identity of sellers buyers and details of prices paid. Its purpose is to guarantee the property ownership identified in the registry. It is also quite reliable, but there are serious access problems both for the public and for municipalities. On payment of a fee an interested party can obtain a copy of a specific deed, but only if he knows the names of the seller/buyer and the code number of the property. This information is not easily obtained by the public.

In theory, this information should be available to municipalities. A 1985 statute introduced a procedure whereby municipalities could claim and receive this information from the finance ministry, but in practice this statute is rarely applied and then only restrictively. Access to this information about the operation of the market would be of considerable interest in analyzing issues related to property and the implementation of land policy.

There is no land register open to the public of the type available in Germany (the *Grundbuch*) providing information on the land itself, servitudes and recent transactions. In three eastern departments, however,(Haut-Rhin, Bas-Rhin, Moselle) there is a *livre foncier* (land register), which is to some extent similar to the German Grundbuch. The Chartered Union of Land Surveyors makes regular appeals for the extension of this system to the rest of the country.

In France, valuation committees of the type known in Germany as *Gutachterausschüsse* do not exist at the local level. Apart from the fiscal assessment (for which a general re-evaluation is now being undertaken), there is only a case-by-case assessment by the finance ministry's *Service des Domaines* whenever a property is being bought by a public authority, for instance by expropriation or by pre-emption. This assessment is obligatory, but the actual compensation paid may substantially vary from the price assessed by the finance ministry (see discussion of expropriation, above).

Sources of information at the national level No statistical processing is done at the national level even though the finance ministry has created a databank (IMO) that gives assessments from 1978 of average prices of various types of land and buildings in all cities with more than 10,000 inhabitants. Average prices are shown in a bracket range resulting from the yearly valuations established *à dire d'experts* (according to the evaluators' experiences in local land and property markets). These local experts use a variety of criteria based on the quality of location (e.g. city centres, and residential and industrial suburbs) and land quality and condition (e.g. whether serviced or not). These criteria do not correspond to any method of zoning. Moreover, IMO valuations may vary significantly from year to year. In fact, the IMO databank makes no reference to categories of buyers or sellers.

Another source of data is the urban affairs ministry. A half-yearly series of data on the acquisition prices of developed land benefiting from a building permit is put out by a subdivision of the ministry known as DAEI. It uses a sample of all building permits for single-family housing that is looked at over time. It is thus possible to study trends by department, region and urban unit. Nevertheless, DAEI provides no data on the volume of market transactions nor on categories of buyers and sellers.

Another database (UTI-SOL) deals with the agricultural land market and is updated annually. The ministry of agriculture uses it for land-use observation. Investigators survey 550,000 or so agricultural areas using aerial photography. A branch of this ministry (SAFER, see Ch. 4.2) conducts data gathering through the *déclaration d'intention d'aliener* (DIA – pre-selling procedure) in rural areas. Two types of information are noted: physical land-use and functional land-use. Information has been published since 1950.

Some local authorities have started to carry out statistical analyses, especially when the DIA is submitted to them. These are, however, specific and cannot be said to be nationally representative. A better tool for observation and analysis of the land market is needed.

At another level, some empirical studies have been carried out by various research groups. These studies are characterized by a qualitative approach, examining the price situation at a given level: city area, labour market, region or department. Also, their sources and methods vary according to their means and objectives: they utilize as primary sources documents such as transfer acts, declaration to sell (DIA in rural areas) and the commercial prices for new subdivisions of family houses. Because of the variety of authorities behind these projects (commune, department, region, etc.) it is impossible to make the resulting data compatible at a national level. Efforts are being made to co-ordinate methods and to set up feasible aims and goals.

Private sources of information Private actors on the market are looking for information at the micro-level: one plot or set of plots of land, one dwelling, etc. Besides the specialized weekly and monthly magazines, which provide information on supply and demand, the trade association of *agents immobiliers* (real-estate agents) has recently created a free fully computerized Minitel service providing information on land supply.

Major developers and builders mainly use their own networks of professional contacts, particularly those concerned with transactions (notaries, real-estate agents and brokers), and, increasingly, elected officials. An interesting new development is the initiative of the notaries' trade association which, thanks to fresh and reliable information, now publishes statistical analyses on the evolution of the housing market in Paris. Unfortunately this is done only for the city of Paris. This could obviously be developed in other areas and also disseminated to the general public, developers and city planners. No information on individual transactions can be released, so the data remain compatible with the principle of confidentiality (on which a high social value is placed in France).

Another private source is the bank Crédit Foncier. This bank specializes in new housing and apartment loans, and has a widely developed network of property experts at the local level. Thus, indirectly, each branch serves as a local source of land data. Since 1990, this bank has started to publish data concerning land and property prices that contain an evaluation of the housing market, the office market and land prices in major agglomerations. However, it has little statistical value at a national level.

Finally, the census provides very useful information on the housing market (but nothing on the land market). It mostly covers property in physical terms (type of dwelling, size, comfort, etc). The results of the last census (1989) are now published.

3.2 The financial environment

The general situation

Questions related to the financial environment, credit practices, restrictions on capital import, etc., are to a very large extent the same for land and property market operations and transactions, and will therefore be further analyzed in Part III.

Investors come from different backgrounds. Since there are no specific restrictions on capital imports, it is impossible to explain and analyze in detail the sources of investment capital. Nevertheless, the question of foreign investment can be approached from two angles: for example, through case studies of industrial development (see Ch. 6.3), and by sectoral study of leisure-related investment (summer residences, especially in semi-rural areas) and tourist-related acquisition.

In rural areas, exhaustive data shows that more and more foreigners who are not farmers are buying property. In total, by 1989 1,700 foreigners had become land-owners in rural France, owning 20,600 ha valued at FF1.5 billion in 1989. This is about 4 per cent of the total number of acquisitions in rural areas. Broadly speaking, most farms acquired by foreigners are located in southern France. The average plot size bought by foreigners is about 12 ha with an average price of FF900,000. Seventy per cent of these new land-owners come from European Community countries, with many of the others coming from Switzerland. Clearly, non-farmers make up the bulk of the total number, especially where sunshine and tourist resorts abound (Roussillon, Perigord, Aveyron, le Lot and the Tarn areas). In the regions bordering the Mediterranean coastline and the Rhône, foreigners purchase between 5 per cent and more than 10 per cent of the land.

Various factors affect financing in each sector: the real interest rate, the economic situation, demand, etc. (see Ch. 1.2). The financing of land acquisition by developers involves supplementary costs loans need to be taken out for longer than for building itself. Two points should be noted: first, banks of different types, including the specialized Crédit Foncier, have developed specific loan structures adapted to different types of land develop-ment. Secondly, a special mechanism (*dation en paiement*) has been intro-duced whereby a developer/builder, instead of borrowing money from a bank to pay for the land from the selling land-owner, "promises" (with the backing of a bank guarantee) to pay in kind. For instance it may provide flats or other buildings instead of money. Thus, the banks can intervene in the process of supplying housing (see Ch. 3.3).

For all investments firms and households can apply to banks for credit, and they can approach specialized financial institutions in accordance with the type of investment or their legal status. Generally, banks' financial resources consist of customers' deposits and those of the specialized institutions are bonds issued on the capital market.

Interest rates for medium- to long-term loans to companies vary, but households usually choose loans with fixed interest rates. For households, except in the case of consumer credit, credit mechanisms are always mortgages on real property or are secured by another property owned by the borrower. A down payment of 20–30 per cent of the investment cost is generally required. For companies, the credit securities required by the financial institutions vary according to their status.

Credit markets are no longer regulated and all agents can find loans at market conditions. Despite this, agriculture and housing still benefit from subsidized government loans. In the industrial sector, only small and medium-size companies can still obtain loans at low interest rates. These are financed by a special subsidy.

Land-banking

All economic agents can finance their land purchases with loans provided to finance buildings, if construction immediately follows land purchase. In other cases the agents can finance land only with short-term loans that generally have to be repaid within two years. The conditions of the loan can vary greatly between different investors.

Financing land-banking The following cases illustrate the financial environment and the rôle of the public sector in particular in the financial aspects of land banking (e.g. Caisse des Depôts & Consignations, national funds for different projects). Generally, the predominant rôle of the finance ministry must be noted because it acts as the allocation referee in cases of large projects (new towns, the La Défense development in Paris, etc.) or specific public organizations (such as the national railways, the armed forces and other ministries) and of local authorities such as municipalities.

For large projects such as land-banking for new towns, a specific fund is created by the competent authority (either as a public or a public/private entity). As stated above (Ch. 3.1), the three examples of public-land operators (AFTRP, EPBS, EPML) also require specific budgets to meet the cost of land, charges for pre-emption and expropriation, inter alia, and for revenue from land resale, added value and any development gains. The EPML and the EPBS's budgets within a defined area are still financed by a special tax, fixed by the advisory committee for each agency. In each zone, the public agency has a ceiling for the total sum to be collected, equivalent today to less than 1 per cent of the four local direct taxes.

For new towns, the state has set up a fund known as the *Fonds National d'Aménagement Foncier Urbain* (FNAFU – the national fund for land and urban development). Until the 1980s, the government's effort had generally been important: FF450 million in the form of budgetary allocations through different state credits (FF550 million from the FNAFU fund). A *Taxe spéciale équipements Savoie* was enacted for the 1992 Winter Olympic Games in Albertville in 1987. This tax applied only to the departments hosting the Olympics and could be used for buying land.

Financing through a ceiling mechanism seems to be adequate and effective as long as these ceiling rules (proportional to a given percentage of local taxes) do not result in a diminishing real value of a specific tax. For example, the EPBS's tax as a proportion of local taxes went from 3.54 per cent to less than 1 per cent within 15 years.

Cases of public administration are more intricate. Since the army and the railway company SNCF are both major land-users, they have a strong leverage position, even though they are not, in legal terms, the land-owners. For example, it has been estimated that the army utilizes 0.45 per cent of the national territory. Since the passage of a 1986 law, 75 per cent of the total revenues from the sale of such land or property was to be returned to the former land-user. The rest of revenues (25 per cent) go to the state budget. Thus, each ministry that owns land attempts to make the most profit out of a sale and it acts as autonomously as possible. Of course, they sometimes barter among themselves: the French army ceded land and property for a new complex by exchanging a Paris site for a site on the outskirts. With a different logic, the SNCF set up a subsidiary organization charged with developing and then selling developed land to developers. If a high added value is obtained, value-added tax must be paid (see Ch. 3.3). Likewise, the army sold an old barracks in Paris in 1989 (in the 15th Arrondissement known as Dupleix) for an estimated FF1.2 billion.

Local authorities have had a freer hand since decentralization. They are able to borrow money from banks and other financial institutions. Until 1975, the municipalities received funds from the FNAFU fund whenever there was an exchange of land or property between administrations. They can also turn to long-term loans from Crédit Local de France (CLF), a bank that specializes in loans to local authorities. Evidently, the state seems to have disengaged from this financing. Today more banks use the findings of credit-rating agencies (e.g. Moody's) to evaluate the local finances for any project, including land-banking. It is worth noting that a reduced number of municipalities engage in land-banking policy.

Insurance companies, some of them public companies, are interested in urban property. They can be instrumental in financing large land-development projects and can finance the purchase of built-on land.

On the private side, the main feature of the recent period has been the radical change in the attitude of developers and builders towards land-banking. This has been essentially because of the rise in interest rates. During the 1960s and at the beginning of the 1970s, the existence of low real interest rates and rapid land-price increases usually made it very profitable for developer-builders to buy land in advance, at a moderate price. They then waited for the development of urban pressures and the construction of the main infrastructure by the public authorities. Moreover, planning and zoning allowed major developers to buy large tracts of agricultural or forest land for long-term development projects.

For private developers this behaviour has progressively changed. During the 1980s, land-banking has increasingly appeared to be a costly device, given the high real interest rates and a changing zoning system. Currently, the increased flexibility of zoning seems to some extent to have enhanced again the practice of private land-banking, but now on a more selective and limited basis than in the 1960s.

Transaction costs

The notion of transaction costs is very complex, since it includes objective measurable costs such as taxation, notary's fees etc., but also various costs that cannot be unequivocally measured, such as the cost of the delay and the cost of obtaining relevant information.

Transaction costs are rather high in France: the taxation system (Ch. 3.3), the information system (Ch. 3.1) and also the behaviour of the actors (Ch. 4.2) confirm this. A recent conference addressed this delicate subject and revealed a lack of ways of assessing transaction costs (ADEF 1991). Reducing the transaction costs, and thereby improving the fluidity of the market, is an explicit objective of the present government. But at the same time, it is necessary to be aware of the possible side-effects of a highly fluid market, which can be prone to speculation. Examples of this increased volatility can be seen, for instance, in the USA.

3.3 The tax and subsidy environment

Taxes and the land market

Table 3.3 summarizes the main taxes having indirect or direct influences on the land and property markets and also lists the relevant legal articles.

The four direct taxes at the local level Four basic local taxes applicable to everyone, except for public administration, the Church, and diplomatic buildings and grounds, form the basis of local taxation (Direction Générale 1991).

Raw land assessed at its rental value by the finance ministry has been taxed for two centuries (Ardant 1971). The tax is in effect mainly a tax on agricultural land (the average effective rate is 0.4 per cent) but it is inadequate for land classified in the local plans as development land, where the effective rate is usually lower than 0.1 per cent.

In 1974 the current system of TFNB came into effect. Since it is a local tax, municipalities are the major recipients: two-thirds of the total tax revenue goes to them. In 1989, FF7,436 million was collected through this tax, of which FF4,500 million went to the local authorities. The raw land tax is based on its rental value with, however, a 20 per cent allowance.

Property tax (TFB) is applicable to all private properties and is also based on rental value. In 1989, FF42,357 million, including FF28 billion paid to local authorities, was collected in property taxes. In total, land taxation yields about FF80 billion per year.

A local dwelling tax is paid by all households, although there are rebates for very low-income families and elderly people. In 1989, the tax yielded FF44,184 million. The fourth local tax, the corporate tax, is levied on trades and businesses, and it generated FF76,002 million in 1989. The wages of a given firm's employees (18 per cent pro rata) are utilized as the tax base.

These four taxes are the core of the local public-financing system since freedom is given to local authorities (by vote) either to raise or lower their rates within brackets determined by the central government's fiscal authorities.

The four direct taxes raised a total of FF169.9 billion in 1989. Table 3.4 shows the range of values of the four local taxes for the year 1990. A wide variety of combinations exists since each local authority can increase or decrease the rates within a reasonable range with the likely effects indicated in Table 3.5.

Table 3.3 Tax environment: synoptic table.

Direct local taxes	Who pays	Target of the tax	Tax base	Law (Articles)	Amount % & FF	Tax allowance	Exceptions	Miscellaneous	Recipient	Revenue in 1987 (in 1989) Bn FF
Raw land tax (TFNB)	land-owners	all except public authorities, diplomats	rental value	1393, 1586 1599bis	0.4% of market price	20%	public lands new: ind, 2yrs dwellings, 10–15yrs	except in IDF collected by fiscal administration	municipalities (2/3) départments & regions (1/3)	6.9 **7.4**
Property tax	property owner	buildings, except public buildings	rental value	1380, 1586 1599bis	varies with municipal-	50%	new: dwellings 10–15 yrs	except IDF cannot exceed 2.5× national average	same as above	34.8 **42.35**
Dwelling tax	occupants owner/tenant		rental value	1407, 1586 1599bis	same as above	rebates for low incomes	varies with household size	50% of households affected	same as above	37.8 **44.1**
Corporate tax (trade & business)	firms, professionals	buildings in relation to wages and plant	rental value, wages, income	1447, 1586, 1599bis	same as above	< or = 5% of added value	max = 5 yrs in declining areas	except IDF cannot exceed 2.5× national average	same as above	61.4 **76**
Wealth tax	house-holds	global wealth	market value of stocks & property	1988 law	0.5% to 1.5%	income tax + wealth tax	circuit breaker	135,000 tax-payers	State	7
Death & gift duties	heirs/ recipients	total inheritance	market value						State	4
Land transfer tax (indirect tax)	buyer	any property exchanged	market value	662, 1594 A art. 35 1985 law	base rate = 13.6 rates vary	reductions: dwelling = 4.2%	purchase by public authorities	Prefect determines rate in departments	mostly départements	20.3

	seller	any property exchanged	deflated capital gains	up to 50% decreasing with time	tax free after 30 years	various deductions	State	not known
Capital gains tax on land and property	seller	any property exchanged	deflated capital gains	up to 50% decreasing with time	tax free after 30 years	various deductions	State	not known
Income tax	household	all revenues	same as above	scaled/ listing	30% allowance on wages & different allowances for various professions	State	260	
VAT on real estate transactions	firms/ individuals	added value	real gains	257.7 CGI 13.02% land 18.6% -	part of VAT recouped by firms		State	NA

Tax movements and their likely impact A given local tax is not permitted to exceed 2.5 times the average national rate. Furthermore, one of the local taxes cannot be raised higher than the pro rata average increase of the other three taxes. For instance, Paris cannot raise its business tax by 5 per cent if the average pro rata increase of the other three local taxes is 4 per cent. Even with these safeguards, tax rates differ widely from one municipality to another. High inequities are due to the unequal distribution of fiscal wealth and differences in financial costs. In the Ile-de-France, for instance, there are significant contrasts between two groups of municipalities, those with the highest rates and those with the lowest rates, as far as land and property taxes are concerned (Tables 3.4, 3.5 and 3.6).

The present system is not satisfactory and reform is needed, as the rateable value of land and property is obscure and often misunderstood by taxpayers. The intricacies of the system, however, make it extremely difficult to revise the system, as Table 3.3 indicates. A reappraisal of this system and its assessment methods is regularly postponed. A percentage of certain of these taxes is used for administrative costs, and the state keeps about 7 per cent.

Table 3.4 Margin of manoeuvre in fiscal pressure (%).

Local tax	Weak value less than	National average	High value greater than
Dwelling	9	11.53	14
Property	11	14.18	17
Raw land	30	36.92	45
Business	10	12.74	16

Source: Guide statistique de la fiscalité directe locale 1990.

Table 3.5 Linkage among the four local taxes.

Tax and its movement		Likely impact
Dwelling tax increases	→	No constraints on other local taxes
Business tax increases	→	No constraints on other local taxes
Dwelling tax decreases	→	Business tax and land tax decrease with no constraints on the property tax
Business tax increases	→	Dwelling tax increases
Land tax increases	→	Dwelling tax increases
Land tax decreases or	→	Business tax decreases with the average
Property tax	→	Rate of the other three taxes

Source: ADEF.

Table 3.6 The highest and lowest land and property tax rates in Ile-de-France (towns with more than 10,000 inhabitants in the Paris region in 1990).

Land[1]	%		%
91 Evry	144.8	92 Neuilly	1.6
91 Etampes	141.3	92 Courbevoie	4.8
91 Vigneux	130.5	92 Montrouge	5.2
91 Courcouronnes	118.0	92 Boulogne	6.3
95 Franconville	117.9	94 Saint Mandé	6.3
78 Trappes	114.7	92 Saint Cloud	6.3
95 Montigny	108.0	92 Issy	8.5
87 Houilles	103.3	78 Le Chesnay	8.7
91 Juvisy	100.7	93 Saint Ouen	8.8
91 Corbeil	100.5	92 Puteaux	9.0
Property[2]	**%**		**%**
93 Villepinte	29.5	92 Neuilly	1.2
93 Villetaneuse	27.5	92 Courbevoie	2.7
91 Juvisy	26.9	92 Boulogne	4.7
95 Osny	26.7	94 Chennevieres	4.9
93 Romainville	25.9	75 Paris	5
95 Cergy	25	78 Le Chesnay	5.4
95 St Ouen L'Aumone	24.2	92 Levallois	5.6
95 Vauréal	23.8	92 Montrouge	6.0
94 Joinville	23.6	92 Puteaux	6.0
93 Le Pré St. Gervais	23.5	92 Saint-Cloud	6.1

Source: ADEF 1991.
Note: National average equals: 1. 37%; 2. 14.18%.

Taxes on speculative profits and capital gains Taxation of speculative profits and capital gains can apply to any private person or company that deals with land development, such as companies dealing in lot subdivision (*lotisseurs*). On rare occasions, firms in other business sectors act as subdividers. The firms that play a major rôle in subdividing on a large scale make this their professional business. They may be stock companies (SA), private limited companies (SARL) or general partnership companies (SNC). The subdivider is solely responsible for any taxes to be paid and any delay in paying them. The general framework is VAT on real-estate transactions (Statute of the 19 July 1976). Table 3.7 shows the general principles of this taxation on land and property, particularly in the cases of subdivisions.

Lotisseurs are supervised by commercial rules governing such things as the minimum number of stockholders and minimum capital, and are taxed on benefits resulting from commercial transactions (i.e. the specific real-estate VAT, which is currently 13.02 per cent). Companies may be taxed up to 50 per cent on their returns. The framework for public-sector developers is different. First, in a normal procedure, their transactions on land are taxable (at 13 per cent VAT) when buying development land (Article 257.7 CGI or General Fiscal Code). In the case of a declaration by a public utility, for instance for the pre-emption or expropriation of land, the municipality does not pay VAT at all (Art. 1042 CGI). If a municipality provides infrastructure or serves as an intermediary, it pays the VAT on the actual price. A return of this VAT might be possible if the condition that a given project of subdivisions comprises more than 10 lots is satisfied (see Rennes case study, Ch. 6.1).

Taxes on urban development The instruments for plan implementation have been stated earlier (Ch. 3.1). Three related tax tools should be mentioned in connection with these: development tax (TLE); the legal density ceiling levy (PLD); and certain payments concerning the floor-area ratio (COS). In addition to these three related taxes, specific charges have been instituted such as the "Olympic Games" tax since 1987 (TSE/Savoie). The case of Ile-de-France has provided opportunities to implement special taxes.

The tax base for local development tax is the construction or reconstruction of all types of buildings. It is based on a lump-sum value of the building project. A scale of rates states the amounts to be paid according to the quality of the building and the net built area. Exemptions are granted to public buildings, foreign embassies and certain projects of public interest. Liability to pay rests with the holder of the building permit. Two taxes complete the TLE device: at the departmental level (environment and open space taxes, near 0.3 per cent) and the Ile de France supplement (1 per cent). Table 3.8 summarizes the proceeds of these various taxes.

About 13,000 local authorities have now benefited from the local development tax. Growth has been positive except in the period 1985–6. Other taxes, such as the betterment levy, have shown the same positive pattern except for the period 1986–7. In general, the figures rise rapidly when dealing with the lesser taxes (i.e. PLD, FAR, etc.) and grow at a relatively constant rate when dealing with the local development tax. If we compare the proceeds of these urban taxes with those of the four local taxes, the contrast is striking: FF2.5 billion and FF169.9 billion, respectively, in 1989.

Table 3.7 Summary of the different taxes for subdivisions.

Duration of land holding	Payments by firms or real estate brokers	Payment by individuals		
		Short term (0–2 years)	Medium term (2–10 years)	Long term (10–30 years)[1]
Calculation of Profits	= sale price minus .acquisition price .charges .diverse works	= sale price minus .value of acquisition .charges or 10% flat rate developing charges	*Non-speculative* transfer price minus .acquisition value revised according to the inflation index .acquisition charges or 10% flat rate revised .developing charges (except financial charges) *Speculative* transfer price minus acquisition value increased by 3%/per year during the first five years and 5% during the remainder .acquisition charges or 10% flat rate increased by 3% to 5% .developing charges increased by 3% to 5%	*Non-speculative* transfer price minus .acquisition value revised according to the inflation index .acquisition charges or 10% flat rate revised .developing charges (except financial charges) *Speculative* the added value sum minus .3.3% per year from the 10th year
Taxation Taxes on firms and benefits	payment in 5 parts	-exception if transfer sum less or equal to FF30,000 -flat abatement of FF6,000 -taxes on added value and benefits payment in 5 parts	-exception if transfer sum less or equal to FF30,000 -flat abatement of FF6,000 payment in 5 parts	-exception if transfer sum less or equal to FF30,000 -flat abatement of FF6,000

Note: 1. For individual holding of more than 30 years no taxes are paid upon realization of the asset.

Table 3.8 Urban development taxes.

Year	Development tax (TLE)[1] 000s	% change	Environment tax (TDCAUE) 000s	% change	Current FF					
					Open zones tax (TDENS)[2] 000s	% change	Levy (PLD) 000s	% change	FAR (COS) 000s	% change
1985	997,537		76,676		161,202		390,450		72,537	
1986	879,311	(11.85)	74,782	(2.47)	145,951	90.45	346,393	(11.28)	108,145	49.9
1987	1,025,463	16.62	112,910	50.99	167,551	114.85	444,392	28.29	77,670	(28.18)
1988	1,081,695	5.48	120,747	6.94	173,819	103.62	474,626	6.8	168,344	116.74
1989	1,218,025	12.6	143,102	18.51	198,272	114.27	719,223	51.53	223,314	32.65

Source: Ministry of urban affairs and fiscal authorities.
Notes: 1. number of local authorities, (municipalities and agglomerations) which have benefited from this development tax: 12,984 (1985), 13,122 (1986), 13,230 (1987), 13,071 (1988), 13,270 (1989).
2. Departmental tax for urbanism and planning for open spaces.

Land-development and related subsidies

Most land- and property-related subsidies are the product of policies and programmes in other sectors (building industries, deindustrialization processes, European Community structural funds, social programmes, etc.). Various administrations can influence the subsidies. A municipality, for instance, can regulate local taxes, for example, by reducing land or property taxes.

At the national level, an inter-ministerial agency takes control of the situation. The government's agency for regional economic development (DATAR), with the help of the EC Commission, can intervene through different frameworks: the *Fonds Interministerial d'Aménagement du Territoire* (FIAT – the inter-ministry fund for planning), *Fonds special des Grands Travaux* (FSGT – special fund for large works), the urban amelioration fund (Banlieue 1989), etc. One of them, the FSGT, is no longer in use. Whenever a region's or agglomeration's economy declines, DATAR is expected to play an active interventionist rôle. This has been the case with land recycling. Two specific aid packages should be noted that apply subsidies to land reconversion. These are state, regional or municipal contracts; and *Primes d'aménagement du territoire* (PAT – grants for regional development).

At the regional level, DATAR co-ordinates and channels the different subsidies, which come principally from the state, the European Community and other administrations, for state, regional or municipal contracts. The EC structural funds seek to promote the development and the structural adjustment of industrialized regions, and thereby enhance the rebirth of these regions. How much of this aid goes to land recycling? DATAR states that in the five years 1984–9, 4,000 ha of brownfield land had been treated (from a total of 16,000 ha in need of treatment), including 1,725 ha still being recycled. It is worth noting that 58 per cent of the surface area treated involves land recycling, while only 4 per cent concerns infrastructure. Areas such as Lorraine, Dunkerque and la Lozère have been supported by different types of subsidies. One of DATAR's aims is to reintroduce new land into the marketplace in the 13 regions in which it funds operations.

Public organizations are in charge of recycling old industrial sites; they then sell the recycled land to prospective buyers, ready for re-use (see Pompey case study, Ch. 6.2). Nationally, during this period, the state has spent FF657,775,624, (nearly 30 per cent of total funding); the EC has contributed FF272,977,691 (12–14 per cent of funds), mainly via the European Regional Development Fund; while other territorial bodies (municipalities, regions, departments) have invested FF1,082,459,375 (53.7 per cent of funding). Breaking up the total figures enables the expenses of recycling, decontamination, and infrastructure provision to be identified. Table 3.8 shows these percentage costs throughout France's brownfield areas.

Table 3.9 Land recycling costs.

Recycling projects in 13 regions	FF in 1988	% of recycling process costs
Demolition/clearing & decontamination	136,308,985	8
Replanting and landscaping	123,829,377	7
Infrastructure	366,000,033	21
Total	626,138,395	36

Source: DATAR 1991.

In addition to these sources of funding, the allowance for regional development (PAT) play a minor rôle in land development, but does mobilize funding to create jobs, to reinforce international firms' location in France and to decentralize economic activities. As with the other funds, DATAR examines and decides which projects should be funded according to its own criteria. The number of funded projects has increased from 65 in 1987 to 115 in 1988, 136 in 1989 and 186 in 1990. Indirectly, it can have an impact on land-use schemes since sums are large: FF255 million in 1987, FF1,023 million in 1988, FF675 million in 1989, and FF834 million in 1990. The three principal regions of benefit were Nord–Pas-de-Calais, the Pays-de-Loire and Lorraine. However, it is still impossible to give precise figures for land development in the cases of PAT allowances.

Property subsidies affecting the land market

The effect of property subsidies on the land market is extremely difficult to identify, since rigorous analysis would require a modelling of the overall tax and subsidy system and of elasticities of supply and demand, and a description of the rôle of financial variables such as the interest rates.

The taxes and subsidies affecting property are described in Ch. 7.2. Some of their effects can be described in a qualitative way. The first conclusion that can be derived from the experience of the past 15 years is that the increase in land values in the peripheral areas has resulted from the household housing subsidy system (APL and *allocation logement*) produced by the 1976 reform and the growth of subsidized loans (*prêts à l'accession à la propriété* – PAP) for the purchase of single-family dwellings (see Ch. 7.2).

This framework has been a powerful incentive for low- to medium-income households to buy houses, given the importance of the subsidy for families with two children or more, and the low interest rates (even lower than inflation in the late 1970s and at the beginning of the 1980s). The increased demand for single-family dwellings has obviously pushed land prices upwards in peripheral areas of major cities, especially in the Paris region. These two factors have been reversed to a large extent: real interest rates have risen (see Ch. 1.2), including subsidized rates such as PAP, and the housing subsidy

(APL) decreases rapidly as soon as children reach the age of 18. The joint effect of these two factors has led some households into a precarious situation, which produced a recent statute on "overindebted households" that attempts to adjust repayment schedules.

Another tax incentive that clearly exerts an influence on the land market is the 1986 statute. This was updated in 1990 and it creates a powerful incentive to rent a house for a period of time: through the deduction of a large part of the investment (FF400,000) of taxable income for households. Jointly, with the relaxation of rent controls since 1985, this incentive has probably had the effect of slackening the decrease in supply of rental dwellings, with an obvious (but non-quantifiable) effect on land values. The fact nevertheless remains that the number of privately owned rented dwellings is still decreasing by 80,000 every year.

In the social-housing sector of the market, the phenomenon that has to be taken into account, especially in major cities and the Paris region, is the influence of land prices on construction costs. The state subsidy for social housing (PLA; see Part III) can be granted only if the price of land remains below a ceiling of FF600–900 per m^2). Local authorities, or private companies through the "1 per cent grant system", can provide special subsidies for the purchase of land whose price is higher, but such subsidies are insufficient in areas such as Paris and the inner suburbs where the price of building land greatly exceeds the ceiling. This phenomenon, explained in Chapter 2, creates a serious problem. The "Delebarre statute", approved in July 1991, was a response to this and it established a "linkage" mechanism through which private builders subsidize the supply of land for social housing in areas where land prices are high.

CHAPTER 4

The process:
the market environment

4.1 Price-setting for urban land

The processes whereby farmland is converted to building land are precisely described in Ch. 4.3, and the different legal procedures that can be used are described: ZACs, subdivisions, simple building permits and land-pooling associations. In addition, an explanation of the behaviour of the main actors in the process is described. During the process, land prices increase from those for "pure" agricultural land (FF1–8 per m²) to fully serviced land ready for construction (from FF100 per m² to FF24,000 per m² in Paris). The mechanism of land-price increases varies in different situations. There are, nevertheless, some general characteristics.

First, it must be emphasized that there is no such thing as the "real" price of land. It is the fundamental result of two factors: the intensity of demand at a given location, and the development rights that are granted by local plans and other building regulations. The evolution of land prices during the development process has changed greatly during the past 15 years, inasmuch as deregulation and the increased flexibility of zoning have increased the phenomenon of "floating value". The price of land does not react in a quasi-mechanical way to re-zoning, but to a large extent it anticipates the change of the local plan. The growing number of "options" in the land market serves to back up this phenomenon. As a result, some prices have gone up to FF100 or even FF200 per m². This can be observed in areas still labelled as agricultural land, so long as there is a strong likelihood that they will be re-zoned in the near future. Developers may also exert some pressure on the elected officials responsible for granting building permits. Broadly speaking, price-setting corresponds to what is known as "backwards accounting", in other words the developer-builder first assesses the demand price for the final product (dwelling, office space), then deducts the construction costs, interest charges, infrastructure costs and other expenses, and then deduces the price that can be paid for raw land (see Fig. 4.1).

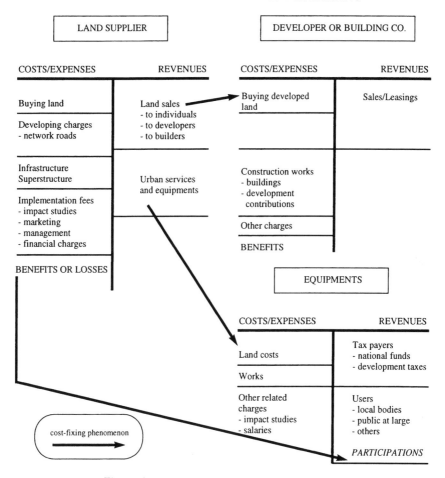

Figure 4.1 Actors in the development process.

An official public valuation of land occurs when a public authority (for instance a municipality) buys a piece of land, either on a voluntary basis or through expropriation or pre-emption. The valuation is then carried out by the finance ministry (see Ch. 3.1) and the price is set (except for pre-emption) at the full market value at the time when the land is purchased. The key question in defining the "true" market price is whether a piece of land can be characterized as land that can be built on; that is, if a building permit will be granted on it.

The definition of "buildable land" has been clarified by a 1985 statute that defines it as a piece of land that is contained in an area that can be developed according to the regulations, and at the same time it must be fully serviced with infrastructure networks adapted to the importance of the construction to be built.

4.2 The actors and their behaviour

The general behaviour of different actors in the land and property markets is discussed here. There are three main groups of actors: public actors at national, regional and local levels; the different intermediaries such as notaries, real estate agents and brokers; and groups of suppliers, especially subdividers and construction companies (Fig. 4.1).

Overlaps within these groups of actors are possible. For example, the national railway company (SNCF) is part of the public domain but sometimes acts with its own private logic. Development and land-banking companies (public or private) can be both suppliers and demanders. Further, most of these public actors have common denominators but may differ according to specific projects and economic goals.

Public actors
Public actors aim to regulate the legal land framework as well as economic conditions, especially for large projects and urbanization schemes (ZACs, new subdivisions, land-banking, new towns, etc.). Utilizing the different urban tools, previously described, the target of the municipalities and other public authorities is to maintain and revitalize the public domain. A public body very often evokes the "public interest" in order to justify its actions and aims. For instance, a ZAC can be presented as a means to attract new activities and jobs in order to fight unemployment and enlarge the tax base. Broadly speaking, a public inquiry determines the necessity and the extent of such actions (see Ch. 3.1).

Another important point is the effect of decentralization on the behaviour of public actors. Municipalities and regional authorities increasingly want to determine what to do and when to do it, without state control. They have signed state contracts (*contrat état/région*, lasting three years) that fix certain aims and goals. The budgetary outlay is included and the public body has the necessary autonomy to carry out its own actions. The financing schemes are very imaginative from this point of view. The local authorities' actions are controlled *a posteriori*; the state has changed its strategy and has gradually become a referee.

When all areas of public or state administration, including the army, the national railways (SNCF), government departments and public utilities, are taken into account, it is clear that public actors are very influential in the land market and subsequently in the property market. The importance of such quasi-public organizations, and the value of their landholdings, is shown in a recent study (Chaline 1987).

Intermediaries

Intermediaries include those who belong to associations or professional bodies (real-estate agents, subdividers, surveyors, etc.) and those who act individually, such as the occasional real-estate broker.

First, the rôle of the public notary in France must be clarified. About 5,000 people enjoy this public status. They fulfil a very important part in the land market as well as in the property market. As stated earlier, the public notaries legalize and transcribe the deed of sale (Ch. 3.1). This deed is then entered into the property register. Another copy goes to the appropriate fiscal authorities. Notaries' fees are about 10 per cent of the total price, including the *publicité foncière* (mortgage publicity). In addition to their official function they may also serve as intermediaries in the buying and selling of transactions. Since they are at the centre of the land and property markets, they are able to assess property and conclude a sale for a third party (a firm or an individual). Other groups of agents are also widely involved in property transactions.

In the group of intermediaries there are also professional trade associations; one important example is the *Fédération Nationale des Agents Immobiliers* (FNAIM – national real-estate federation). Created in 1946, FNAIM co-ordinates more than 6,500 members nationwide. Although the majority are real-estate agents, members include about 1,100 real-estate brokers. Its aims are to represent and to defend the interests of its members. To coordinate its proposed services, such as information systems, FNAIM organizes conferences in order to enhance the economic and social rôle of real-estate professionals. The association enables its members to provide better services and advice (through legal, economic and administrative counselling) to their clients. The FNAIM is represented in 21 out of the 22 regions of France and it claims that its members conclude 300,000 land and property transactions per year with a total value of FF100 billion. According to FNAIM, this represents 45 per cent of total transfers. Association officials say that approximately one in six contracts is advised and helped by this network. Because of its importance as a lobby group, FNAIM debates with public authorities (law-makers, state and municipalities).

Real-estate brokers are typical intermediaries, buying and selling any type of property on their own account. Most of them have to be placed in a special category, for two main reasons: they are in a no-man's land, institutionally speaking, and, since they are both buyers and intermediaries, their rôle is ambiguous. The result has been abuse of the system. They are required to register with the *Tribunal de Commerce* (register of commerce) so that they fall into a special category in terms of fiscal status (see Ch. 3.1).

A recent assessment of real-estate brokers concludes that it is impossible to calculate their exact number (estimates range from 4,000 to 17,000).

Certainly there are at least 4,000 professionals who profess to be real-estate brokers. As stated earlier, they have played a major rôle in booming areas, especially the Ile-de-France. For instance, in a sample of the transactions in 1988 in central Paris (Paris intramuros), 1,000 out of 7,000 transfers were concluded by brokers (ADEF 1991). Their aims are to buy and resell land or properties after refurbishment or in the same state (taking high risks and contributing to the boom spiral) (see Ch. 5.3 and 5.4). Apart from their activities inside cities, they can also be, for example, subdividers, developers and contractors in semi-urban areas (Ch. 3.1).

Private land suppliers

On the supply side, site subdividers and developers can be part of the *Syndicat National des Aménageurs Lotisseurs* (SNAL – the national organization of subdividers and site developers). There is much diversity in this group: some are full-time professionals, while others practise only occasionally. Subdividers prepare and sometimes develop their plots, but as a rule they undertake only one part of the whole subdivision process. This lobbying organization of *aménageurs* and *lotisseurs* has 136 company and 350 individual members. It aims to promote the interest of its members at all levels and emphasizes the fact that its members contribute over a quarter of development land each year. All its members utilize detailed knowledge of the local economic and political contexts.

In legal terms, subdividers can take advantage of different legal statuses. Broadly speaking, if the goal of the subdivider is not an immediate development and construction project, he or she can choose to operate as a real-estate broker. This enables avoidance of transfer rights taxation when selling unprepared land. Only VAT on gross margins must be paid if the subdivider sells unprepared land within a five-year period. Often a subdivider postpones construction by up to nine years. If he decides to go ahead with a land-development scheme, he then pays the 13% VAT on developed land retrospectively (see Ch. 3.1). The largest subdividing company (Foncier-conseil, with 17 offices nationwide) has an average output of 1,500 to 2,000 plots per year. The second largest (Devicq-France Lots) contributes half that number.

Promoters and builders can take the initiative and the financial risk by constructing dwellings or apartments. This group, especially the builders, should not be confused with subdividers or developers, or even with investors (who maintain ownership of the resulting product). Once a subdivision of developed land has been bought, the promoter has to get financial support from banks, and technical advice from architects and economists, and must begin marketing. In view of the economic uncertainty, promoters depend more and more on large civil engineering groups, such as Bouyges, and large financial institutions such as Crédit Foncier and BNP. They have been progressively absorbed to become part of the vertical organization of the land development and construction industry through a series of subsidiaries (see also Ch. 8.2).

Land surveyors

Land surveyors are grouped into the *Ordre des Géometres Experts*, a nationwide chartered organization. Since 1947, they have been in charge of delineating and measuring private property boundaries. Thus, their expert rôle is to guarantee the accuracy of land and urban property measurements. The surveyors' justification is the vital importance of marking clear boundaries and dimensions to property to be sold or exchanged. They also help the fiscal interests (concomitant to the cadastre's use); land surveyors ensure that the eventual sellers are responsible for their acts (Article 1602 and 1603 of the Civil Code). Most subdivision measurements are done by these surveyors. In 1989, the surveyors' organization signed an accord with SNAL agreeing to the professional responsibility and liability of the surveyor for the accuracy of dimensions of properties sold by SNAL members.

Levels of training and public relations

Levels of education, training and self-regulation by the different actors discussed above fall into two main groups. The first consists of public sector and technical professionals; the other group is made up of professionals from the private sector.

Public organization and administration has a specific regime (see Part I). Most of the agents belong to the state and other public bodies. Policy-makers are supposed to represent the public interest and they are judged on this basis. Notaries fall into a special category because they are a form of scribe. Chartered surveyors and other professionals – architects, urban planners, economists, sociologists, etc. – are required to go through special higher education programmes that usually last four years. Each profession has its chartered organization, which is responsible to its members for training and public relations.

The other group is made up of real-estate agents and brokers, and financial engineers; they can enter land and property affairs either by practice (three or more years) or by training (two years minimum at university). In addition, each profession may have specific academic courses designed to train future professionals.

In the fields of advertising and public relations, all actors have developed a very important strategy. They follow the same trends as in other economic sectors. Using the same public-relations media (TV, newspapers, etc.), these lobby groups transmit their point of view as regularly as possible. Moreover, they organize annual conferences and seminars in order to inform their respective members. The majority of them have a bulletin. Brochures, advertising and press conferences are part of this strategy. To illustrate this point, public notaries of the Ile-de-France and a loan bank (Crédit Foncier) publicly announce their annual data, consisting, respectively, of details of mortgages concluded in the Ile-de-France, and lending patterns and prices. Public administration and other public agencies generally appear to be following suit. Nevertheless, some well known intermediaries, notably in the commercial and building sectors, are cautious about diffusion of their data and they charge a high price for it.

Reasons for the behaviour of these actors

Does the legal and economic framework affect each actor's behaviour? Each group of actors modifies its strategy in line with the changing parameters (economic and legal among others). Financial incentives may add to or restrain the output of land supply (see Ch. 5). In a gloomy economic environment, lobby groups (builders, subdividers) usually ask the government to lower the relevant taxes. Speaking in 1991, the president of the national federation of builders (*Fédération National des Promoteurs Constructeurs*, FNPC) argued that the law (LOV), then in draft, would mean higher prices for dwellings.

Equally important have been the legal changes that have affected the behaviour of each group of actors (see Ch. 3.1). Municipalities and other territorial entities have engaged in methods other than rigid zoning. They have attempted to attract new businesses and entice new industries (where possible those that create less pollution) by both fiscal and urban measures. Pre-emption tools, ZACs and other useful urban devices allow public actors to adapt to new economic and industrial parameters. The tools for infrastructure provision give the municipalities a margin to negotiate with private partners in the land-development process (i.e. PAE, TLE; see Ch. 3.1 and 3.3).

4.3 The process of plan implementation

This section presents a brief overview of the different legal–economic devices through which an area or a piece of land can be developed: subdivision, comprehensive development schemes (ZAC), land-pooling associations and the direct building permit.

The building permit

The most straightforward development procedure is as follows: a land-owner, whether an individual or a building firm, applies to the municipality for a permit, assuming that a local plan exists. This building permit must correspond to the specifications of the local plan or ZAC project in respect of the type and the size of the building, floor area ratio, etc., and any relevant protection plans such as environmental or historic site plans. The municipal authority will decide whether to give explicit approval or refusal. At times, a delayed authorization (*sursis à statuer*) can be issued in order that further documents may be consulted. Public authorities must state their reasons for approval or refusal in one of three circumstances: delayed authorization, rejection of the proposed scheme (in order to guarantee effective judicial review later on), or approval of the proposal with amendments.

Two procedures are prescribed once a building permit has been granted, so that neighbours or concerned citizens may object to a given permit. To inform third parties, the building permit must be displayed at the town hall (public posting). This notice should remain for two months after the date on which authorization was issued (in case of judicial review by the administrative courts or public inquirer). This notice does not offer technical information, but all concerned citizens may have access to the available documents. Secondly, a building notice must be displayed throughout the whole construction period on the plot where the future building(s) will be constructed. Authorities are particularly stringent with builders or companies that overlook this formality.

Subdivision (lotissement)

Whenever a landholding is divided into more than two plots for development, an *autorisation de lotissement* (specific subdivision permit) is required. The *lotisseur* (developer) who owns the land, often after having assembled several neighbouring parcels, services the whole area with infrastructure. He then divides it and resells plots for single-family dwellings with the guarantee of obtaining a building permit for the eventual buyer of the plot (see Rennes case study, Ch. 6.1). Since initial investment is minimal, most private lotisseurs operate with limited resources and are very responsive to the economic situation. Although the majority of subdivisions are done by private entities, public authorities can utilize the subdivision procedure. There are two categories of lotisseurs: occasional and professional (see Ch. 3.2 and 3.3).

Zones d'aménagement concerté
(ZAC – comprehensive development schemes)

Created by the 1967 LOF law, the ZAC process of plan implementation affects land expropriated or pre-empted by local authorities in order to provide it with infrastructure and other services. The ZAC process, designed for large development schemes, requires two stages of authorization. The scheme's coherence is established by a *plan d'aménagement de zone* (PAZ – specific development plan) and an infrastructure programme.

Steps for ZAC schemes If a local plan exists, the municipality first delineates the ZAC's limits and objectives. If this is not the case, and if more than one municipality is involved, the prefect delineates and fixes the ZAC's goals and aims. Secondly, an impact study or file must be prepared describing the development plan and detailing technical elements such as the height of buildings, roads, sewerage lines, green spaces, etc. Thirdly, a given ZAC project is presented to the local citizens, for it is then ready to be implemented once the relevant public authorities approve it. This is the same procedure as for local plan approval (see Ch. 3.1). Once approved, it supersedes and overrules the local plan in the delimited area. De facto, the local plan is modified to include the ZAC prescriptions if they differ widely from those in the local plan. Fourthly, a more detailed ZAC project study is prepared, including provisional costs for infrastructure and the schedule for the timetable of development and financial outlay (see ZAC Citroën case study, Ch. 10.2). Fifthly, the ZAC is submitted to a public inquiry that precedes the DUP (see Ch. 3.1). Finally, the ZAC must be approved by the competent authorities: the prefect or municipality or both.

This procedure can be applied to any type of development or redevelopment (new peripheral development, urban renewal, industrial zones or shopping centres). A ZAC can be located only inside areas zoned "urban"(U) or "development" (NA) in a local plan. In the absence of a POS, a ZAC can be designated only if there is a master plan, with which it is compatible. Practically, it seems that the ZAC procedure is not so different from the German *Bebauungsplan*. The case studies of Alsace (Ch. 6.3) and ZAC Citroën (Ch. 10.2) illustrate how the public enquiry and expropriation process takes place within the ZAC procedure.

Public or private ZACs Afterwards, operations may be, and often are, delegated to the private sector. Of course the public authorities, which initiated the land development, can set out a *cahier de charges* (guidelines) specifying binding contracts with private developers. They can specify, for example, the number of road access points and the green area. A distinction must be drawn between public ZACs, in which the municipality or a semi-public company retain, the financial responsibility for the development, and private ZACs, for which a contract is reached between the public authority and a property developer. The latter is responsible for implementing the scheme, and taking the risks and profits. The Citroën-Cevennes case study (Ch. 10.2) presents a typical example of a public ZAC.

Land-pooling associations

In the procedure involving an *Association Foncière Urbaine* (AFU – land-pooling association), several neighbouring land-owners wish to develop their land on their own, but the land-ownership pattern necessitates replotting of the area before development can begin. It must be authorized by the prefect. The AFU process allows the land-owners to reallocate and service their land. They are allowed to expropriate the reluctant land-owners, within a delineated area, subject to the approval of two-thirds of the land-owners,' whose holding must also amount to at least two-thirds of the delineated land area.

The AFU enjoys the legal status of an *établissement public* (public body) and collects development fees from its members. Rarely used, partly because of its legal complexity, the process is however a promising one since it widens the possibility of replotting and at the same time offers the municipality a private supply of public works.

As a whole, the various tools that have been briefly described in this section cover a wide range of situations corresponding to various types of public–private partnerships, different sizes of development schemes and different development and redevelopment objectives. But on the precise point of implementation, there is probably a missing link, in the form of either an obligation or a strong incentive, to promote development according to the content of the plans.

CHAPTER 5

The outcome of
the urban land market

5.1 Changes in the structure of ownership

There is no national survey that covers the structure of land-ownership. Nevertheless, some major trends may be identified. Within urban areas, full ownership of land and property has decreased, to be replaced to a large extent by *copropriété* (co-ownership). This mostly applies to multi-storey buildings formerly owned by a single owner who rented the apartments. For various reasons, but primarily because of decreasing returns from renting housing, these buildings are being sold flat by flat to be transformed into co-ownership. The legal status of co-ownership then encompasses the surrounding land, which becomes the common parts of the co-owned property. This evolution has been especially rapid in Paris (Table 5.1).

Table 5.1 Structure of ownership in Paris.

	Number of buildings put into co-ownership	Percentage of the total	Number of co-owners
1935	2,600	2.1	5,710
1950	5,998	3.5	5,998
1985	57,735	48	841,000

Source: Patrice de Monean, *A qui appartient Paris?* Editions Seesam, 1987.

As regards suburban land classified by local plans as development land, a survey completed in 1988 on a representative sample of 50 municipalities provides information on the structure of land-ownership and a typology of land-owners (Table 5.2)

Table 5.3 indicates that, despite the large number of small plots in suburban areas, it is clear that large land-owners still own a large part of land in such areas. Tables 5.4 and 5.5 indicate overall ownership by age of land-owner and type of ownership by age of land-owner.

A final trend identified is that land-banking by public authorities, which was an active policy of both municipalities and specialized public land banks in the late 1960s and 1970s, has decreased, but the trend now seems to be reversing.

Table 5.2 Who owns development areas? (NA zones of local plans).

Type	Number (%)	Area (%)
Households	89	64
Public authorities	5	19
Property companies	4	8
Other companies	2	8

Table 5.3 Size of the land property in which the parcel is included.

Size (m^2)	Number (%)	Area (%)
0–2,000	30	3
2,000–10,000	27	11
10,000 +	43	86

Table 5.4 Age of landowners.

Age	Number (%)	Area (%)
0–39	17	10
40–59	37	36
60–79	39	44
80 +	7	10

Table 5.5 Type of property by age of owner.

Age	Natural areas	Built-on property	Development land
	%	%	%
< 30	2	0	6
30–39	7	26	40
40–49	13	29	42
50–59	22	17	8
60–69	26	19	3
70–79	20	4	1
> 80	10	5	0

5.2 Demand for building land and its supply

The increased flexibility of zoning has progressively modified the structure of demand and supply by increasing the demand for agricultural land located in peripheral areas of major cities from non-agricultural users that anticipate re-zoning.

Finally, Table 5.6 summarizes the evolution of land-use since 1975. During this period the total area of artificial and built-upon land has grown from 5.3 per cent to 6.8 per cent of the total land area; it can be seen that this growth corresponds mostly to a decline in cultivated areas. By comparison, woods and forests have grown from 27 per cent to 27.6 per cent.

Table 5.6 Land-use by area.

Category	1975 Area	%	1980 Area	%	1985 Area	%	1986 Area	%
Woods & forest	14,828	27.0	14,863	27.1	15,127	27.5	15,140	27.6
Cultivated areas	31,963	58.2	31,525	57.4	31,048	56.7	31,009	56.4
Other natural areas	5,037	9.2	6,035	9.2	4,941	9.0	4,923	9.0
Artificialized and built-up areas	2,912	5.3	3,365	6.3	3,665	6.7	3,707	6.8
Total	54,919	100	54,919	100	54,919	100	54,919	100

5.3 Prices

Differences exist in the evolution of prices because of spatial reasons and because of the structure of each sector (plots for one-family housing, urban land for industrial and commercial purposes, etc.). Relative land prices (in current French francs) have varied widely between Paris and the provinces: a ratio of 4.9 in 1978, 10.2 in 1985 and 16.5 in 1989. By comparison, other ratios between major cities excluding Paris are more stable: 0.97 in 1978, 1.52 in 1985 and 1.45 in 1989. The ratios in small cities range from 1.02 in 1978 to 0.76 in 1989.

The impact of spatial differences on certain markets
In spite of the dualism between the Paris region (Ile-de-France) and the 11 other cities described in Ch. 1.5, and the land price trends (see Ch. 1.4), the price contrasts are remarkable, especially in Paris after the historic trebling of prices in the period 1986–8 (Granelle & Guelton 1991). The trend was also witnessed in other areas but with less force. Price signals in 1989 and 1990 disclosed a levelling of prices, in particular in Paris. In the following paragraphs, data on urban unit prices and the prices in 50 cities, for city-centre and peripheral locations, are compared.

In the period 1976–86, price patterns differed according to urban size and service provision. Even though there are limitations to the statistical data available (see Ch. 3.1), it is worthwhile noting the evolution of the predominant prices between the maximum and the minimum, for example, the growing gap between the Ile-de-France and the rest of the country, even though there was a halt in 1984–6. Prices in effect followed the economic situation of the time: a recession and a deflationary period led to lower prices, especially in 1984–6. Accordingly, land prices in rural areas declined by 5.8 per cent in constant francs (1986). In urban units of less than FF100,000 land prices went down by 3.5 per cent. Urban units between FF100,000 and FF200,000 saw land prices decrease by 9.8 per cent, while for urban units in the range FF200,000 to FF1.5 million land prices went down by 10.5 per cent. Finally, even in Ile-de-France land prices declined by 8.6 per cent (EF 1986, 1988).

Land prices in France between 1978 and 1988 are shown in a more constant form in constant values in Table 5.7.

Table 5.7 Sample data on the evolution of prices of developable land.

	FF/m^2 (1989 prices)					
	1978		1983		1988	
	Price	No of	Price	No of	Price	No of
Paris[1] (7 peripheral arrondissements)	4,679	119	5,519	52	14,272	128
Paris periphery (6 locations)	681	223	753	149	1,044	167
Centres of large cities (10 villes)	754	209	942	200	1,540	357
Peripheries[2] of large cities (18 locations)	301	2,725	385	3,036	467	4,979
Centres of medium-size cities (12 cities)	703	89	561	68	850	219
Periphery[2] of medium-size cities (35 locations)	199	4,624	229	2,939	249	3,759

Source: Etudes Foncières 45, December 1990.
Notes: 1. data for the arrondissements 11, 12, 13, 14, 18, 19, 20. The most expensive arrondissements in Paris are not included. 2. Definition of ordinary periphery according to IMO data.

Nevertheless, the growth rate in central areas of the 10 major cities followed a contrasting pattern. Prices have increased more rapidly in downtown areas than in the outskirts. The particular case of Paris intramuros shows a growth of 159 per cent during the 1983–8 period. By contrast, the Paris outskirts saw land prices rise by 39 per cent in this period. Concerning the price trends between peripheries of large cities and those of medium-size cities, land prices have increased between 8 and 21 per cent. The rate of growth of downtown areas in both types of cities has been much the same: 63 per cent and 51 per cent, respectively.

Prices per sector (plots in ZAC and subdivision process)
The prices of individual plots for single-family dwellings also differ according to the region. Table 5.8 shows for the end of 1988, the beginning of a boom period, the cheaper and more expensive land markets. Overall, the average price for an average-size lot of 914m^2 was FF172,000 and regional average prices vary between FF99,000 and FF346,000. Ile-de-France,

81

Provence/Côte d'Azur, Rhône–Alpes and Alsace seem to maintain their values whereas Limousin, Franche-Comté, Lorraine continue to be disfavoured.

Table 5.8 House plot sizes and prices.

Regions	Average size (m²)	Average price (FF 000s)	Price per m² (FF)
Ile de france	687	346	503
Provence – Côte d'Azur	1,073	246	229
Rhône–Alps	984	202	205
Alsace	814	190	233
Picardie	832	157	189
Languedoc – Roussillon	691	154	223
Nord – Pas de Calais	877	150	171
Pays de Loire	862	136	155
Champagne–Ardennes	938	135	144
Aquitaine	1,109	131	118
Haute-Normandie	893	129	145
Midi–Pyrénées	1,129	129	114
Centre	1,016	129	127
Bretagne	906	126	140
Basse-Normandie	1,056	119	113
Bourgogne	887	118	133
Poitou–Charentes	1,030	112	109
Auvergne	815	111	137
Lorraine	756	111	146
Franche-Comté	920	105	114
Limousin	1,235	99	80
TOTAL FRANCE	914	172	118

Source: DAEI, December 1988 prices.

As far as price trends of individual plots in ZAC or subdivisions and individual plots in diffused areas are concerned, a 1989 survey, monitored by the urban affairs ministry, shows that land prices are generally higher in ZAC or subdivision areas than in diffused areas (Table 5.9). This is true even if all categories of urban units are taken into account. In rural areas, for instance, a ZAC/subdivision plot costs FF140 per m², whereas a standard plot costs FF71 per m² in diffused development areas (Table 5.10).

The average plot size has been reduced Compared to 1975 figures, from the 700–1500 m² bracket to 500–550 m². The latest figures currently show a stable trend in land prices for individual plots throughout the country (Tables 5.12 and 5.13). The average price is FF200,000 to FF300,000 per lot. Prices are more homogeneous and relatively comparable to the 1989 prices, especially in provincial cities. In Ile-de-France, prices varied from FF250,000 to

FF2,000,000 per plot in 1990. Broadly speaking, after a boom period of three years, land prices have stabilized.

Table 5.9 Prices of individual plots in diffused areas.

Size of urban units	Area average (m²)	Price average (FF)	Price (FF per m²)
Rural communes	1,459	104,000	71
Urban units (inhabitants)			
at least 20,000	1,273	129,000	101
20,000–100,000	1,102	147,000	133
100,000–1,500,000	1,198	209,000	174
Ile-de-France (outskirts)	519	441,000	713

Source: DAEI survey, June 1989 prices.

Table 5.10 Prices of individual lots in ZAC and subdivisions.

Size of urban units	Area average (m²)	Price average FF	Price FF per m²
Rural communes	919	128,000	140
Urban units (inhabitants)			
at least 20,000	723	152,000	211
20,000–100,000	668	154,000	231
100,000–1,500,000	711	168,000	290
Ile-de-France (periphery)	488	351,000	645

Source: DAEI survey, June 1989 prices.

Table 5.11 Lot sizes and prices

Surface area (m²)	Number of observations			Average prices (FF000s)		
	ZAC and subdivision	Diffused	Total	ZAC and subdivision	Diffused	Total
< 400	179	51	230	158	224	173
400–500	300	61	361	176	229	185
500–600	334	69	403	172	183	174
600–700	274	59	333	162	159	161
700–800	214	63	277	177	187	180
800–1,000	220	107	327	189	172	183
1,000–1,200	137	155	292	179	145	161
1,200–1,500	62	125	187	187	145	159
>1,500	100	287	387	237	140	165
Total	1,820	977	2,797	177	162	172

Source: DAEI, Survey June 1989 of prices of December 1988.

In Tables 5.12 and 5.13, five types of markets can be discerned. First, markets in recession characterized by low prices, a decline in demand, an increase in supply or both. The cities of Lorient, Mezières, Besançon and

Belfort are in this situation. Secondly, heterogeneous markets are character-ized by an array of different situations between the centre and outskirts. This occurs, for example, in important cities such as Bayonne, Compiegne and Toulon. Thirdly, unbalanced markets where demand has increased but supply remains at low levels, for example, in Strasbourg, Aix-en-Provence, Reims, Lille, Grenoble. Fourthly, markets characterized by their stability and an abundance of land. Finally, the sustained or growing markets are character-ized by relatively higher prices, in particular in Dijon, Amiens, Poitiers and Le Mans.

Prices in the development process

Using the example of a nationwide developer, Foncier Conseil, a cost:sale ratio in the land-development process can be observed. Cost:sale prices clearly differ in relation to the cities (Table 5.14).

Table 5.14 also shows the current acquisition prices for developed plots, averaging FF500–550 per m². Although these are indicative land prices, there is again a widening gap between the Ile-de-France and the rest of the country. Thus, in Versailles (Ile-de-France) plots cost FF446,000; by contrast, prices oscillate around FF175,000 per plot in the Lorraine area and go up to FF192,000 per plot in the Nantes area. On the whole the rate of price growth in large agglomerations has been estimated at 11.2 per cent during the 1989–90 period, having been 10.6 per cent in the preceding period. In conclusion, many land-market analysts agree that a market transformation has taken place, revealing various "micro-markets".

Prices in the industrial property market

There is also a widening gap among regions in the industrial property market. The considerable and general increase in land prices is linked with the evolution of the office market. It should be noted that 1989 was characterized by high records and general euphoria in the property market. The industrial property market in the 17 agglomerations surveyed by the Auguste-Thouard Group is considered to be a neglected market. Prices rose by 10.6 per cent in 1988–9. Three major agglomerations lead in price increases: Lyon, Marseille and Lille. The 1989 prices are the result of a constant rise in prices that started in the mid-1980s. Although there are differences between cities, Table 5.15 shows the average values for various cities.

Table **5.12** Land prices for clustered dwellings.

City	Dominant prices(FF) C	P	City	Dominant prices(FF) C	P
Neuilly, Boulogne, Levallots	10,000		Val-de-Marne (Nord Est)	8,000	
Puteaux, Suresnes, Courbevote	7,000		Yvelines (Secteur 1)[1]	7,000	
Nice	5,000	2,800	Hautes-de-Seine (Grpe 3)	5,000	
Montmorency (Canton et Vallee)	4,500		Cergy-Pontoise	4,000	
Strasbourg	3,600	1,600	Lyon	3,500	1,800
Annency	3,200	1,700	Reims	3,000	1,500
Vallees Bievre et Yvette	3,000		Nantes	3,000	1,200
Aix-en-Provence	3,000	1,800	Arrondissement Argenteuil	2,800	
Avignon	2,600	2,000	Seine-Saint-Denis[2]		2,500
Toulouse	2,500	1,300	Angers		2,500
Roissy, Ozoir, Pontault	2,300		Marseille	2,300	1,600
Essone (Val D'Yerres)	2,200		Marne-la-Vallee		2,200
Le Mans	2,200		Bordeaux	2,000	1,500
Yvelines[3]	2,000		Plaine Gonesse, Z' Roissy	2,000	
Vexin francais et Normand	2,000		Fontainebleau, Melun		2,000
Essonne (Nord, Val D'Orge	2,000		Orleans	2,000	1,000
Tours	2,000	1,300	Compiegne	2,000	1,300
Yvelines[4]	1,800		Melun-Senart		1,800
Grenoble	1,800	1,400	Nancy	1,800	1,100
Montauban	1,800		Meaux		1,800
Montpellier	1,800	1,400	Bayonne	1,700	900
Rouen	1,700	1,000	Metz	1,700	1,200
Mulhouse	1,600	1,300	Chartres	1,600	1,200
Vannes	1,600	900	Laval	1,600	
Colmar	1,500	1,100	Seine-Saint-Denis[5]	1,500	
Evry	1,500		Pau	1,500	800
Saint-Quentin-en-Yvelines	1,500		Chambery	1,500	1,300
Lille	1,400	1,200	Caen	1,350	1,000
Brive	1,300	500	Nimes	1,200	1,000
Troyes	1,200	600	Seine-Saint-Denis[6]	1,200	
Dijon	1,200	1,000	Calais	1,200	
Saint-Etienne	1,100	700	Angouleme	1,100	
Le Havre	1,100		Lorient	1,050	400
Cherbourg	1,000	700	Besancon	1,000	
Evreux	1,000		Valence	1,000	
Amiens	1,000	600	Nevers	900	800
Limoges	900	600	Auxerre		

Notes: C = city centre; P = periphery.
1. Arrondissements de Versailles, Saint-Germain en Laye. 2. Courbon, Villepinte, Tremblay, Gagny, Neuilly-sur-Marne, Neuilly-Plaisance, Le Raincy, Villemonble, Pavillons-sous-Bois, Livry-Gargan, Aulnay-sous-Bois, Sevran, Clichy-sous-Bois, Montfermeil, Vaujours. 3. Arrondissement de Rambouillet. 4. Vallee de la Seine, Arrondissements de Mantes-la-Jolie, (Asineres, Bois-colombes, Clichy, La Garenne-Colombes, Colombes, Vill-la-Garenne). 5. Saint-Ouen, Saint-Denis, Aubervilliers, Pantin, Pre-Saint-Gervais, Lilas, Bagnolet, Montreuil. 6. Ile-Saint-Denis, Epinay, La Courneuve, Bobigny, Villetaneuse, Stains, Pierrefitte, Dugny, Le Bourget, Drancy, Blanc-Mesnil, Bondy, Noisy-le-Sec, Romainville.

Table 5.13 Price per plot for single family dwelling.

City	C 000s	P FF	City	C 000s	P FF
Puteaux,Suresnes, Courbevoie		1,200	Hautes-de-Seine(Groupe 3)		1,200
Yvelines (Secteur 1)[1]		1,200	Cergy-Pontoise		800
Annecy		750	Nice	1,500	750
Val de Marne (Ouest)		700	Seine-Saint Denis[2]		600
Val de Marne (Est)		600	Plaine Gonesse, Zone de Roissy		600
Seine-Saint-Denis[3]		500	Essone (Val D'Yerres)		550
Valles Bievre et Yvette		500	Essone (Nord, D'Orge)		500
Saint-Quentin-en-Yvelines		500	Toulon		450
Avignon	350	450	Yvelines[4]		450
Grenoble		450	Roissy, Ozoir, Pontault		450
Lyon		450	Marne-la-Vallee		450
Evry		450	Yvelines[5]		400
Vexin Francais et Normand		400	Fontainebleau, Melun		400
Essone (centre)		400	Melun-Senart		380
Aix-en-Provence		350	Marseille		350
Bayonne	500	350	Etampes, Ferte-Allais, Milly		320
Meaux		320	Strasbourg		300
Dijon	360	290	Montpellier		290
Pau	350	250	Chambery	300	250
Rouen		250	Colmar		250
Rennes		250	Nantes	300	230
Le Havre		230	Lille		230
Nimes		220	Mulhouse		220
Metz	260	220	Bordeaux		215
Vannes	360	210	Orleans	250	210
Toulouse		210	Compiegne		210
Saint-Etienne	350	200	Perpignan		200
Nancy	250	190	Chartres		190
Brest		190	Clermond-Ferrand	330	180
Besancon	230	180	Belfort	220	180
Tours	200	180	Evreux	250	160
Amiens		160	Beauvais		160
Caen	205	155	Le Mans	240	150
Arras		150	Lorient	200	150
La Rochelle	190	150	Auxerre	180	150
Angers	160	150	Saint-Brieuc		150
Calais		150	Bourges		150
Macon	180	140	Montauban	160	130
Poitiers	160	130	Charleville–Mezieres		130
Carcassonne		120	Moulins		120
Laval	150	110	Cherbourg	140	110
Limoges	180	100	Nevers	150	100
Brive	160	90	Angouleme	140	90

Notes: C = city centre; P = periphery.
1. Arrondissements de Versailles, Saint-Germain en Laye. 2. Courbon, Villepinte, Tremblay, Gagny, Neuilly-sur-Marne, Neuilly-Plaisance, Le Raincy, Villemonble, Pavillons-sous-Bois, Livry-Gargan, Aulnay-sous-Bois, Sevran, Clichy-sous-Bois, Montfermeil, Vaujours. 3. Ile-Saint-Denis, Epinay, La Courneuve, Bobigny, Villetaneuse, Stains, Pierrefitte, Dugny, Le Bourget, Drancy, Blanc-Mesnil, Bondy, Noisy-le-Sec, Romainville. 4. Vallee de la Seine, Arrondissements de Mantes-la-Jolie, (Asineres, Bois-colombes, Clichy, La Garenne-Colombes, Colombes, Vill-la-Garenne). 5. Arrondissement de Rambouillet.

Table 5.14 Costs and prices of developed land.

	Cost per operation	Sale per[1] operation	Operations	Number of plots	Cost per plot	Sale per plot
	1990 figures (000s FF)					
Dijon	6,451	37,974	3	211	30.57	180
Nancy	1,681	6,328	2	36	46.75	175.7
Colmar	6,684	18,492	3	76	87.47	243.3
Orleans	13,245	47,119	4	251	52.77	187.7
Nantes	4,765	17,288	3	90	52.94	192
Rouen	5,007	15,017	4	74	33.34	202.9
Versailles	9,198	18,302	4	41	224.34	446.4
Melun	14,296	44,383	1	130	109.97	341.4
Epinay	42,296	150,961	5	476	28	317
Bordeaux	4,946	12,513	3	71	69.66	176.2
Toulouse	1,838	6,079	1	31	59.29	196
Lyon	2,840	5,731	3	25	113.6	229.2
Montpellier	10,535	32,893	3	137	76.9	240

Source: Foncier Conseil and ADEF 1990. *Note:* 1. Sales activity of Foncier Conseil.

Table 5.15 Prices and rents of industrial buildings.

	Lille	Lyon	Marseille
	1989 prices (excluding VAT)		
Old			
Downtown rents	200	200–250	280
Downtown sales	1,800	2,000–3,000	1,800
Peripheral rents	150	150–200	180
Peripheral sales	1,000	1,400–2,000	1,600
New			
Downtown rents	280	na	350
Downtown sales	3,000	na	3,200
Peripheral rents	220–250	240–300	330
Peripheral sales	1,800	1,800–2,800	2,700
Top values			
Rents	400	na	300
Sales	3,500	na	3,500

Source: Auguste-Thouard 1990.

Furthermore, factors such as inadequate supply (85 per cent of the buildings are old) and high demand in peripheral areas (95 per cent of the total), can be added to the economic situation. In 1990, the industrial property market showed many signs of weakness, mainly because of low supply. The industrial property market works more and more on the basis of a triangle: investor–user–developer. Developers are searching for builders or investors so that this mechanism can work and the triangle can be completed. Investors are attracted by the profitability ratio, especially in boom areas such as western Paris. They compare the yield for industrial and commercial development to returns in housing or other sectors and the comparative advantages within each agglomeration.

Investors have recently been participating less and less in the industrial sector. As a result, the developers have adjusted their supply by slowing down project completion rates. This is especially true in the office sector; down by 24 per cent since the 1989 boom.

5.4 Land speculation

There is an ambiguity in the definition of speculation. The speculator can be described as an "economically rational actor" forecasting accurately, and thus improving the economic efficiency of the market in accordance with neoclassical economic theory. He can also be described as a harmful actor using any loophole in the legislation to reap windfall benefits that often result from public works or chancy re-zoning. Another distinction can be drawn between passive speculators, who are waiting because they do not need cash at the moment and think that property is a good hedge against inflation, and active operators in the market. The passive attitude is obviously widespread, and the evolution of the urban land markets over the past 30 years confirms the validity of this choice.

Active speculators are looking for windfall gains on land (or buildings), and take into account any relevant information that will allow them to purchase or sell at the right moment. Active speculation, by individuals or companies, has clearly been increasing in France, especially since the 1987–90 property boom. This is so despite the high transaction costs (see Ch. 3.2). Its growing importance can result in a "bubble" phenomenon, where the price of property does not result from fundamental factors such as GNP and demography, but from the hope of capital gains.

The phenomenon has most dramatically been observed in Tokyo, where the price of land in central areas is now disconnected from its use value. The present situation in Paris or the French Riviera cannot be compared to that prevailing in Tokyo, but nevertheless seems to include some elements of the "bubble" effect.

CHAPTER 6
Case studies

6.1 Provision of developable land in the Rennes District: two subdivision cases

Introduction

The authorities of the Rennes Agglomeration – the regional authorities and the *District Urbain de l'Agglomération de Rennes* (DUAR – Rennes District) – have had a strong land-banking policy since the immediate postwar period. This case study presents and analyzes the influence of a private and a public developer in preparing land for development in the district of Rennes. Statistical data and primary sources are exploited, in particular over the past five years.

By looking at the supply side, two examples of *lotissement* (subdivisions), one public (in the municipality of Bruz) and the other private (in the municipality of Pace), will uncover the complexities of land preparation in the case of Rennes. Knowing that both actors are vital in the local process of subdivision, the advantages or disadvantages of each in terms of cost, lack of comprehensive urbanization process, etc., should become apparent.

The aim of this study is to examine how each actor proceeds in the conversion of raw land into developed land in the Rennes District. The first part will set the local and regional contexts of urban and economic growth and the rôle of administrative and economic partners, in particular the rôle of the *Agence de l'Urbanisme et de Développement Intercommunal de l'Agglomération Rennaise* (AUDIAR) in urban planning. Land-banking policy and its effects will be briefly explained, and relevant economic and social data will be presented. The second part identifies and describes the background of each subdivision. This part will define their locations in relation to the Rennes District. Their respective goals and the schedules of development will be presented and an assessment of the concrete achievements will be put forward both in urban design and financial terms. The third part will take a closer look at the real timetable (i.e. in the face of obstacles of every nature) of both operations; and the financial and economic partnership at local and regional

89

level will be evaluated. Finally, the results of both projects will be compared in order to discern the local land markets.

In conclusion, influential factors and the structure of the land market will be discussed in relation to the comprehensive changes in land policy and urban planning presently occurring in the Rennes District.

Local and regional contexts

The geographic and economic background Situated in the west of France, the region of Bretagne has a population of almost 3 million (2,793,300 in 1988) in its four departments. One of these, Ille et Vilaine, is one of the most prosperous parts of Bretagne and had 786,900 inhabitants in 1988. Eleven per cent of the economically active population is employed in the primary sector, 20.3 per cent in the secondary sector, 7.3 per cent in construction and 61 per cent in services. The traditional urban centre of Rennes competes with Nantes and Brest to be the region's leading city.

The Rennes District (DUAR) is at the centre of Ille et Vilaine. At present it comprises 28 municipalities. Clearly, the city of Rennes is predominant in demographic and economic terms in DUAR.

Table 6.1 Demography in Rennes.

	Number of municipalities	Population (000)	Population growth		Area[1]	Density[2]
			1968–75 (%)	1975–82 (%)		
City of Rennes	1	190	9.5	(1.9)	50.5	3770
DUAR	28	310	16.9	6.95	492	590
Hinter-land	122	410	16.4	27.5	2422	170

Source: INSEE 1990.
Notes: 1. km^2 covered; 2. Density per km^2.

Throughout the 1960s, 1970s and 1980s, the agglomeration of Rennes underwent rapid urban and economic change. Historically, the rise and fall of industrial activities in the agglomeration of Rennes followed after the 1950s. In the 1962–75 period, major industrial development was located here. This contrasted with Rennes' past since this city and its surroundings had been relatively untouched by the industrial boom until 1945. An historian labelled 19th-century and early 20th-century Rennes as the "parasitic city" (Denis 1973).

After 1945, various factors, such as the insight of the mayor, H. Freville, and decentralization (instigated by DATAR) aimed at moving industrial activities from the Paris region to provincial centres, helped stimulate the creation and migration of firms. The 1975–85 period was marked by an economic downturn. The two oil crises of 1973 and 1979, the freezing of industrial decentralization by DATAR, and technological innovations, have changed industrial location patterns. Many old industrial firms went bankrupt at the end of the 1980s. Consequently, unemployment grew and a recession hit the economy of the Rennes District.

Despite this, the agglomeration has remained dynamic. Innovative activities (e.g. a technological park), the adaptation of the remaining industrial firms and the attraction of tertiary firms are tangible results of a coherent economic policy at the agglomeration level. Public and private efforts have paid off, notably the development plan for the Rennes agglomeration (*Plan de Développement de l'Agglomération Rennaise*, PDAR). The transportation, distribution, and service sectors are today among the most dynamic in terms of employment, growth and economic expansion. The employment structure reflects this, showing growth in the tertiary sector and decline in the primary and secondary sectors. The unemployment problem has receded and the area has recovered in the rapid growth years of the 1980s.

At the departmental level, the Rennes agglomeration has fulfilled a very important function in the regional economy. Because of decentralization policies since the 1970s, industries (Citroën, Antar, Thomson) and some public administration offices have been located in the agglomeration. For example the *Ecole Nationale Supérieure des Télécommunications*, SUPELEC (the national electronic school, and research labs at the university and branches of the *Centre National de la Recherche Scientifique* – CNRS) have been relocated there. Today, economic activity is centred on mechanical engineering, automobiles, publishing, electronics, computers and telecommunications production and research.

A recent detailed study (Guengant 1989) classified the DUAR's urban growth into five phases (Guengant 1989):

○ 1960–64 – feeble and even negative peripheral growth; the suburban communities had lost population and only the centre of the DUAR (within the perimeter of the city of Rennes) had gained inhabitants.

○ 1965–9 – a transitional phase; the peripheral communes started to experience population growth. This can be regarded as the suburban take-off phase.

○ 1970–74 – an acceleration of urban development; this coincided with deregulation in the real estate sector that favoured housing supply. Meanwhile, invigorated housing demand resulted from the implementation of the saving and loan housing plan (see Ch. 2).

○ 1975–80 – reversal of the preceding urban development trends; the two oil crises had three side-effects: housing starts plunged, the real-estate market became inert, and construction delays were frequent.

○ Since 1981 – an improvement in the economic situation; it even recovered from the critical 1980 situation after the second oil shock. As of today, an economic upturn has regenerated this urban development, but the uncertainty of the current economic situation and changing patterns of housing needs (more clustered units rather than single-family dwellings) leave many unanswered questions about spatial growth patterns.

Actors and institutions at different levels Throughout the 1950s and 1960s the state played a major rôle, through its usual technical structures and relay institutions, in reconstruction (1946–55) and urban planning efforts in the region. The state took a centralized approach until the early 1970s. After this, DATAR, the agency for regional and economic development, advocated decentralization and put it into effect in terms of industry and economic activities. This agency studied, advised and proposed the location of various economic activities and public administrations in the region. Along with the central government efforts, private organizations such as the chambers of commerce and industry and (*Comité du Développement Economique et Social du Pays Rennes*, CODESPAR – the committee for social and economic development) have served as catalysts of land development.

In order to reconstruct and develop the Bretagne region (including Ille et Vilaine), in October 1957 various governmental branches joined local efforts by creating an action group, the *Société d'Etude pour l'Aménagement et l'Equipement de la Bretagne* (SEAB) in October 1957. It regrouped the *Caisse des Dépôts et Consignations* for financing projects and a development agency (*Société Centrale de l'Equipement du Territoire*, SCET) for the location of economic infrastructures. Two years later, a public–private company, the *Société d'Economie Mixte pour l'Aménagement et l'Equipement de la Bretagne* (SEMAEB), succeeded the old SEAB but had the same rôle: to study and implement policy on economic development at the regional level. In addition, its semi-private status enabled local banks and other actors to participate. It was particularly active in developing policies for land-use, infrastructure, tourism and industrial development. Until 1970 the city of Rennes effectively delegated its land-banking to SEMAEB. After this period, the DUAR assumed responsibility at all levels.

DUAR became a legal entity on 9 July 1970, acquiring a new dimension. Created as a public body, composed of 28 neighbouring municipalities, it has embodied specific aims and goals regarding planning schemes and the implementation of infrastructure and urban servicing. Once two-thirds of the communes concerned, representing more than 50 per cent of the total population, had chosen this alternative, the prefect of Ille et Vilaine issued an ordinance defining the area and the district's headquarters. This is a typical case of municipal co-operation.

DUAR's composition and rules were set up by the representative members.

One of its most important prerogatives is the levying of taxes to pay for urban services. It fixes, collects and manages the charges pertaining to urban services (water supply, sewerage systems and mass transportation) as well as housing and fire department services. Other tasks include economic development and a comprehensive planning scheme for the whole district.

The establishment of DUAR brought about changes in urban policy. Since its creation in 1972, AUDIAR has extended its enabling powers on urban development. In fact, it has come to be regarded as the executive body for urban planning and land-use issues. The urban affairs ministry has progressively let the Rennes District take over this rôle since the beginning of the 1970s. Before 1970, the *Direction Départementale d'Equipement* (DDE) of the urban affairs ministry was responsible for the preparation of local and structure plans. From May 1972 onwards, the newly created AUDIAR was charged with preparing and controlling the urban and planning policies implemented by the District, the member municipalities and the state. Encouraged by the political authorities, the DDE and then AUDIAR implemented a strong land-banking policy until the end of the 1970s, in the same style as that employed in Sweden.

In the 1980s, because of incipient industrial decentralization and the implementation of cohesive development, AUDIAR executed by delegation the conception, regulation, programming, financing, land policy-making, realization, management and all follow-up urban and economic matters. This included its land-banking policy.

Land supply policy: land-banking by the DUAR

Precedents Land-banking has had a rich history since the postwar years in the Rennes agglomeration. Even before becoming a legal structure, the agglomeration of Rennes was influenced by a strong public policy during the postwar period. After 1947, a socialist-inclined municipal team was incumbent in Rennes city hall and influenced local political life until the 1970s. Consequently, an unchanging policy on land and urban development prevailed in city hall. It aimed at continuing the reconstruction effort and repairing the war-damaged infrastructure (water supply, sewerage, etc.).

Tools, objectives and master plan implementation The legal tools utilized for land-banking purposes – expropriation, pre-emption rights and compulsory purchase for public authorities at the local level – have been described in Ch. 3.1. The 1958 law served as a point of reference and enabled the municipalities to utilize compulsory purchase orders to carry out land-banking policy. As a rule, this has seldom been used comprehensively at the national level, but Rennes is exceptional since public authorities there have implemented a sort of "Swedish Policy". In other words, they initially bought raw land at low prices (which they rented to farmers when it was not needed). As they progressively sold the land, according to the demand level, the land-banking policy curbed the price hikes of land and permitted the benefits of land-value appreciation to be shared. From 1973, SEMAEB and the state were ready to continue forceful land-banking, on condition that the District contributed to the funding. Sites were targeted in Bruz, Cesson and Thorign.

The global cost was estimated in 1974 at FF25 million for the next five years. Funding for the transitional stage came from the *Caisse d'Aide à l'Equipement des Collectivités Locales* (CAECL – the national fund for local authorities) and the *Fonds National d'Aménagement Foncier Urbain* (FNAFU – national fund for urban and regional support which each supplied 50 per cent. More CAECL loans followed for specific projects, as well as subsidies and loans from the Caisse des Dépôts et Consignations.

This land-banking policy has three clear objectives: to control the housing market, to control through economies of scale the global costs of urbanization, and to cross-subsidize the urban growth costs for infrastructure and superstructure (ADEF 1990). In this context, subdivision has been a main public policy tool in the DUAR jurisdiction for two reasons: its flexibility in human and technical resources, and its reproducibility. Its flexibility enables the subdivision to have a diversity in architectural and economic terms. The principle of reproducibility means that a given operation can be multiplied a number of times using the same legal and financial mechanisms, resulting in an urbanization of five or 10 subdivisions, provided that demand exists.

To put these goals and aims into practice, the master plan was implemented in 1973. Looking ahead to 2010, it assumed a total population of 250,000 inhabitants in Rennes, 70,000 in the southwest (Bruz, Chartres, Chavagne), 90,000 in the northeast (Cesson, Thorigné) and 140,000 in the other 19 municipalities. In 1975, the local plans of the majority of the DUAR's municipalities showed a remarkable conformity with the master plan as implemented. According to these, it was estimated in 1982 that 2,200 new dwellings would be needed to meet the demand for the coming years. In 1988, issued building permits showed a strong sectorization of the land market: 87.7 per cent of the DUAR's clustered dwellings took place in the city of Rennes; 86.6 per cent of one-family housing in the other municipalities. Within the one-family housing market 70 per cent of the building was done by individual owners. The gap in plot area between public and private subdivisions has decreased.

In addition, a trend of returning to the agglomeration's centre (Rennes city) has taken place, encouraging urban imbalance and segregation, yet, because of the subdivisions, plot prices are lower than in other agglomerations.

Table 6.2 Average surface of
plots in the DUAR (m²).

	Municipal	Private
1979	637	771
1980	641	944
1981	295	822
1982	518	739
1983	833	681
1984	494	704
1985	499	876
1986	513	548
1987	527	545
1988	514	542
1989	518	512

Source: AUDIAR.

Land prices in the DUAR

Constant land prices seem to be the main characteristic of the DUAR's price evolution. Fragmented data show that, throughout the 1960s, 1970s and 1980s, land prices remained stable. In the period 1954–66, the case of Rennes was already considered exceptional since real prices of land had maintained the same level (Granelle 1970). Real prices in 1962 appeared to be the lowest of the period, followed by a slight increase in 1964.

Throughout the 1970s and the 1980s, parallel evolution took place: land prices remained stable mainly because of the policy of land-banking. In 1990, two sources stated that land prices in the Rennes area had levelled (see Ch. 5.3). While this may be true, price differences appear among subdivisions because of two factors: the private or public rôle in developing land, and conditions and structure of timetables (costs, financial charges, commercialization, effective demand).

A thorough study in 1985 indicated the price trends in the subdivision's market. The effective share of the subdivision market between public and private origin showed two tendencies: public subdivisions had grown from 1964 to 1976 in proportion to the number of municipalities that engaged in this type of operation. By contrast, private subdivisions declined steadily from 1964 to 1968 in comparison to the number of municipalities. It then remained stable for almost a decade (1967–76), declined slightly until 1982, and has since stabilized.

The latest price trends are as follows. Plot prices for new housing have stabilized during the past four years. Even though land prices per plot depend on the geographical location of the municipality and the level of provisions, housing loans (PAP, PLA, etc., see Part II) have decreased in volume. This has resulted in the decline of housing supply and therefore a standstill in demand for land. In the second half of 1988, the average price per plot, whether private or public, was FF171,382 in the DUAR; in the first half of 1989 this went up slightly (by 2.19 per cent to FF175,382 per plot).

To obtain an optimum balance between public and private activity, the sectors co-operated in the production of building land throughout the major urbanization period of 1960–85. The difference in land supply has, in effect, been institutionally based, because of a land-banking policy that resulted in a quasi-monopoly of raw land supply. Nevertheless, vigorous and direct competition between private and public developers has taken place.

This sharing has been a clear characteristic of the land market in the DUAR. Despite this, the public sector has largely dominated, with a 70 per cent market share at times. In addition, public authorities have reduced the number of subdivision projects and introduced other innovative procedures such as the ZAC (see Ch. 4.3). In brief, these ZAC procedures have taken the place of municipal subdivisions, yet more mixed and private projects have appeared. As a result, the share of the private sector increased significantly between 1978 and 1985.

Another factor concerns the cost:sale ratio in both municipal and private subdivisions. Table 6.3 shows the average costs for private and municipal subdivisions.

Table 6.3 Average costs of municipal and private subdivisions.

	Average costs			
	Municipal subdivision		Private subdivision	
	FF/m^2	%	FF/m^2	%
Infrastructures/networks	101.1	70	107.74	50
Land acquisition	28.6	20	35.55	16.5
Financial fees	14.3	10	72.19	33.5
TOTAL	143	100	215.48	100

Subdivisions in the DUAR

There is a direct relationship between land supply and individual housing starts in the DUAR, characterized by the distribution of housing starts. Table 6.4 shows that during the 1984–8 period the volumes in Rennes city and the periphery are almost equivalent. Nevertheless, in the following period increased construction took place in the city of Rennes.

In addition the rate of sales has increased from 20.2 per cent in 1983 to 30.9 per cent in 1988 and 45 per cent in 1989. This implies that the property market (and hence the land market) has seen rapid growth in sales transactions.

Table 6.4 Average housing starts in Rennes.

	Rennes city	Annual averages DAUR Rennes	Total DAUR
1978–80	755	1,061	1,816
1979–81	895	1,044	1,939
1980–82	875	1,033	1,908
1981–83	884	1,049	1,933
1982–84	719	1,047	1,766
1983–85	808	1,045	1,853
1984–86	836	966	1,802
1985–87	1,222	1,034	2,256
1988	2,310	968	3,278
1989	2,185	1,006	3,196

During the past 20 years public interventions curtailed land pressures through a comprehensive land-banking policy. The AUDIAR bought land at fair prices. Then it provided the necessary infrastructure and later undertook construction in the case of *lotissements communaux* (public subdivisions), or sold the raw land to private firms for preparation and subdivision.

Legal changes have altered the rôle of developers in Rennes. Before 1960, private subdividers were almost unconstrained by regulations. Today, by contrast, they must submit their projects to the local authorities, who exercise more power of decision than before, and a *cahier des charges* (specification document) is required.

In the case of Rennes, both the public and private sectors have co-existed, fulfilling their respective rôles. This co-existence has helped the production of suitable land for construction. Various factors explain this co-existence from the political perspective: public authorities may influence land supply and control the urbanization process, and undo the errors of preceding flawed private subdivisions.

The preparation of land for development in the DUAR: two case studies
The two case studies discussed in this section are the municipalities of Bruz (with 8,018 inhabitants) and Pace (with 5,656 inhabitants). They are both in the DUAR, within a 15 km radius of Rennes city centre. They also have comparable rates of infrastructure provision: 86 per cent for Bruz and 74 per cent for Pace, and both have surface areas in the 30–35 km² bracket. Finally, the ratio of active population to total population is more or less the same: 43 per cent and 46 per cent, respectively. They differ in that Bruz, 12 km south of Rennes city, had a better fiscal base in the 1980s, and has had a lower rate of annual population growth (0.98 per cent) than Pace's (4.29 per cent).

The genesis of the two subdivisions The subdivision project in Pace is one of the most recent in the area. The subdivision of the Domaine du Bois de Champagne has a total surface area of 140 ha. At the start, 5 large plots were reallocated by a single developer. This subdivision is located in a developing zoning (NA). It is a predominantly residential area, mainly made up of one-family homes. Some observers claim that this development resembles the American model. In 1990, the project was already occupied, the DDE Office (Ministry of Urban Affairs Departmental Office) having followed up the development stages of this subdivision. Equipment taxes and other servitudes were also applicable.

The subdivision of Bruz studied here, Vau Gaillard, is located to the northeast of the municipality. The perimeter of this project is surrounded by a wooded area to the north: a future sport and leisure area. The genesis of this subdivision was in March 1986. The zoning by the local plan (POS) indicated urban zoning (U). At the initial stage, a FAR of 0.6 was granted. With respect to infrastructure charges, the municipality decided to exclude this subdivision from development tax (TLE) since it was the community that was to be charged for this project. The municipality engaged in a process of infrastructure provision estimated at FF8,220,000 in 1986.

The aims of the two projects The municipal subdivision project of Vau Gaillard in Bruz, which resulted in the construction of 350 one-family houses had three main objectives: to fulfil the demand for new housing both within Bruz and in surrounding areas, to realize a planning and infrastructure programme, and to conclude the radial–concentric development of this municipality.

By contrast, the subdivision of the Domaine du Bois de Champagne in Pace was a private venture with different development objectives in two respects: it is a low-scale project providing mostly one-family housing (145 plots) with only three plots planned for multi-family housing, the details of which are to be designed later; and it has a flexible character since there is no fixed FAR or other type of constraint other than those of the local plan.

Outcomes of subdivision in the DUAR

The timetable of operations In Pace, three sections of the Domaine du Bois de Champagne subdivision were programmed, and project implementation was divided into two phases:
First phase
○ general ground-levelling
○ infrastructure provision (sewerage, drainage, etc.)
○ provisional access road

○ primary network connections (phone, water, electricity, cables)
○ secondary network connections to houses
○ commissioning of all urban services.
Second phase
○ finishing off of surrounding areas (pavements, roads, asphalt, etc.)

Table 6.5 Timetable of operations in Rennes.

	Plot numbers	Total number of plots	Average surface (m^2)
1st section	1–50	50	505.4
2nd section	51–110	60	482.5
3rd section	111–166	55	417.3
Other	166–168	3	2980.0

If the marketing process and sales go well, these stages of the timetable must be started within eight months of the approval of the project by the competent authorities (DDE and municipality). All works related to this subdivision must be completed within three years of the initiation of the first stage; and other sections of the project within six years.

The Vau Gaillard subdivision in Bruz is part of the municipality's comprehensive urban programme. This programme consists of a new cemetery, a community centre, a habitat project of 100 dwellings (known as La Noë) near the centre of Bruz, a sports complex, and renovation of the city centre. The Vau Gaillard subdivision project was presented as a major part of this urban programme, comprising the following building programme:
Global programme (target 328–363 dwellings)
○ construction by building companies of clustered housing (from 70–115 single-family houses or 110 mixed single and multiple-family houses,
○ construction of 110 single-family houses by individuals
○ construction of 24 dwellings by individuals or by promoters/builders (depending on the market).

The initial project timetable stated that this subdivision would take three or four years to finish. It will include:
First section (eastern side)
○ construction by builders of 60 single-family or intermediary dwellings
○ construction by individuals of 90 single-family dwellings
○ construction either by builders or individuals of 17 single-family housing dwellings.

Assessment of concrete results Data on the outcome of the Pace subdivision have unfortunately not been issued, but a summary of prices and sizes of parcels up to 1990 is available. Development in the municipality of Pace happens to be dominated by the private sector. Plot prices and other data of interest are shown for the 1985–90 in Table 6.6.

Table 6.6 Land prices in Pace.

	Number of plots	Average price FF/m^2	Average value of plot (FF)	Average plot surface
1/7/1985	5	255	247,350	970
1/1/1986	3	255	272,850	1,070
1/7/1986	8	322	218,638	679
1/1/1987	1	394	146,000	370
1/12/1987	21	377	165,476	438
1/6/1988	9	399	159,333	399
1/12/1988	10	294	200,600	683
1/6/1989	3	270	235,000	872
1/12/1989	2	273	235,000	860
1/6/90[1]	15	361	154,400	428

Note: 1. No private subdivisions; results of public subdivisions which have occurred for the first time in many years.

Table 6.7 Land prices and cost estimates.

	Cost including various taxes 1986 FF	Percentage of total cost
Infrastructure works	31,000,000	70.3
Land-related charges	3,320,000	7.4
Honoraries, fees	3,355,000	7.5
Other expenses, financial, etc.	6,500,000	14.6
TOTAL	44,475,000	100.0

Table 6.8 Evolution of land prices in Bruz.

Date	Number of plots	Average price FF/m^2	Average value of plot (FF)	Average plot surface
1/1/87[1]	21	232	195,112	841
1/7/87[2]	140	296	194,780	656
1/12/87[3]	108/130	345	183,259	530
1/6/88	61/73	345	193,836	561
1/12/88	48/64	346	203,325	588
1/6/89	44/59	346	205,555	594
1/12/89	32/44	346	215,978	624
1/6/90	71/79	346	223,755	647

Notes: 1. Figures include only a small private developers sales; 2. Includes a few plots from private developers; 3. Municipal plots and total plots.

A provisional budget had been drawn up for the Bruz municipal subdivision. Table 6.7 shows the estimates of different costs in relation to the subdivision in Vau Gaillard.

Table 6.8 shows the evolution of prices and the number of parcels for sale from 1987 to 1990 for Bruz.

Conclusion

Although it is rare for public or semi-public sector bodies (e.g. SEM, OPHLM, SIVOM) to undertake land assembly and preparation for subsequent development themselves, the case of the DUAR is an exception. Its public policy demonstrates how a grouping of local authorities under the umbrella organization of the AUDIAR agency have dealt with this issue in a voluntary manner. On behalf of the 28 communes, DUAR (through the AUDIAR) has enacted an active policy of land assembly and preparation for development (in the case of public subdivisions). On the other hand, AUDIAR has allowed (and encouraged) private subdividers to prepare land for development. Consequently, private subdivisions have sprawled in the agglomeration.

The decline in the number of plots for sale in the past five years has had several effects. There has been an adjustment to the demand curve, since the competition among municipalities has become fierce. In addition, infrastructure costs show that the municipal subdivision foresaw the costs and implemented diverse mechanisms in order to satisfy future needs. Some parts of the schedule have not been completed to the original timetable because of the economic situation.

By examining two examples of subdivisions, we discover that the average private subdivision in Pace is marginally more expensive than the average municipal one (Tables 6.6, 6.7 and 6.8). Private plot prices also follow the same price trend as other comparative private subdivisions. There is, in effect, a stronger fluctuation of prices in relation to market conditions within the private subdivision, with prices ranging from FF255 per m^2 to FF394 per m^2. In addition, the private subdivision has benefited from the location of a new secondary school and a sports complex which bolster effective demand. The two secondary schools will be funded by the state, the region, and the DUAR. As far as is known, there has been no direct contribution for these superstructures from the developer except for the usual development taxes. In this sense, the private project gains from these particular public projects.

On the municipal side, prices have remained stable because of the disparity of the cost structure of subdivisions and the infrastructure and superstructure expenses that have been foreseen. AUDIAR is important in this process since it cross-subsidizes the efforts and manages a very important land asset.

But a double market phenomenon appears, since private developers suffer from a handicap when compared with public developers, who benefit from the very low input price of raw land. Land was bought long ago at a reasonable price, as agricultural land, and this has definitely affected the cost structure of subdivisions in general and in particular those in the public sphere, i.e. AUDIAR's. This generally leads to a paradox in land-banking policy. The private developers' view is that this mechanism needs to be thoroughly evaluated, and some criticize its inherent bias.

Furthermore, attitudes towards subdivision in the DUAR are changing because of two structural changes: the patterns of decentralization call into question the whole basis of inter-municipal co-operation, regardless of what AUDIAR's official say or do; and the ideological view of controlling urban growth through land-banking is changing. DUAR's aims and goals also place emphasis on the ecological aspects of economic development. There are, therefore, several issues to be resolved.

6.2 The reconversion of the steelworks site of Pompey (Nancy)

Introduction

France, like many other European countries, has experienced uneven development: currently ranging from the most dynamic regions, such as the Ile-de-France and the southeast, to the regions in decline such as the north and northeast (see Ch. 1.5). Lorraine in the northeast is a typical old industrial region that has faced a significant recession and a decline in its steel, coal and textile industries.

Some of Lorraine's economic areas have unemployment rates of 15–20 per cent, and population, with the Limousin, is in decline. As regards its economic structure, a decrease in the secondary sector of industry has been observed and this loss has not been compensated for. Urban and social difficulties have gradually appeared.

Because of their importance, the steel areas have perhaps been the worst hit by this general evolution. Jobs in the steel industry have decreased from 100,000 in 1960 to just 18,000 today. And urban population has dwindled by more than 20 per cent in 10 years. The result is that local authorities must solve their huge problems with decreased financial means. The large number of municipalities and various political difficulties have hindered real co-operation.

The phenomenon of mono-industry is well known: economy, population and territory are structured by and for the main activity. Its disappearance or decline causes a national problem that must be treated comprehensively. Public authorities have attempted to reverse this massive decline since 1984, aiming to invigorate the economic environment.

The first stage of recycling, beginning in 1984, was characterized by a willingness to attract new firms using substantial fiscal and other incentives. The land and the buildings abandoned by the steel industry were put on the market without improvement and at very low prices. Sometimes, land and buildings were given away by the steel industry under pressure from local authorities. This practice had negative effects on the local land market: land depreciation and competition between these spontaneously created industrial zones and pre-existing ones led to lower prices. Revenues from the transfer of land were lower than the recycling operation costs, with the resulting deficit being borne by local authorities. The firms attracted by this situation were often unreliable, and many have since disappeared.

These experiences explain why public authorities changed their way of working at the end of the 1980s. Since then, more ambitious policies have been carried out with various aims: to change the structure of local space, the local economy and the training process for workers (DATAR 1990a). Since 1986 the first priority has been recycling of derelict land. The policy aims to acquire sites, upgrade their environment, and offer the cleared sites for economic and urban redevelopment. The case study of Pompey shows the reasons for this policy, the methods used and the results.

The context

Pompey is one of two steel areas located near Nancy, most of steel industry being in the northern part of Lorraine. Founded in 1872, the Pompey plant was continuously developed up to 1975, but it was closed in 1986 because of adverse national and international economic circumstances. In 1975, the local area provided 5,000 jobs in steel manufacture and 3,700 jobs in mechanical engineering and metal industries. By 1990, the steelworks had disappeared, and around 1,000 jobs in mechanical engineering and metal industries were left. The number of employed people in the zone has decreased from 21,500 in 1975 to 17,500 in 1990 (INSEE 1986). At the same time, the population has decreased by about 2,000 in Pompey, but has increased in the municipalities closer to Nancy. At the moment, the population of the five municipalities in the area is about 30,000.

The area, which during the steel industry years was economically and culturally autonomous from Nancy (300,000 inhabitants) at a distance of 10 km, is becoming an outer suburb of Nancy, but its lack of residential attractiveness does not generate much real-estate development. In Nancy, as in other parts of France, there is a migration to the centre of towns and a decline in peripheral residential growth, especially in economically hard-hit areas.

The geography of the valley in which Pompey is situated itself limits development possibilities. The plateau is composed of woods, completely protected by the master plan (SD) and local plans (POS), and the sides of the valley are already built-up or in agricultural use and hence are protected. Since the valley bottom was used by the steel industry, large sites were created that are now empty and derelict.

These factors explain why the land market in Pompey is so undynamic, especially for housing, where there has been almost no new construction for several years. Interestingly, price levels in real estate are similar to those in the first belt near Nancy. This may indicate that, if land becomes available, the market will be boosted.

The industrial land market is in a different situation (Table 6.9). The main zones around Nancy have been developed to the south and the west of the agglomeration. At the moment, 100 ha is available in the agglomeration, and annual consumption is about 20 ha (Auguste-Thouard 1990). But several projects are planned. The consumption of industrial land in the area of Pompey is about 3–5 ha per year, with a supply limited in the existing zone to a former steel site. An important reserve (80 ha) was created by the closure of the steel plant, and the commune of Custines intends to create an agricultural zone of 20 ha.

In addition, the location of the area, 10 km from Nancy and 50 km from Metz, is potentially attractive since it is crossed by the A31 autoroute (Luxembourg—Nancy—Lyon) and by an international railway line. But without available land and the infrastructure to link the town to the autoroute, this positive location is yet to generate any real change.

The economic shifts in the area during the past decade have also led to a recasting of the actors facing the recycling issue. After the closure of the steel plant, the public authorities drew up a plan to help create new jobs on the site of the former factory. Now, after demolition of the main buildings and superstructures, the site has been sold to the *Etablissement Public de la Métropole Lorraine* (EPML – public land agency for the Lorraine metropolitan area), which was created in 1973 to conduct land and real-estate deals for public purposes. This has occurred within the framework of the regional policy for Lorraine, which has aimed to recycle all derelict land within Lorraine with the help of the state, the regional council and the European Community (see Chs 3.1 and 3.3).

The three main local authorities concerned (Pompey with 6,000 inhabitants, Frouard with 8,000 and Custines with 3,000) face financial difficulties because of reductions in their tax base. This is especially true of Frouard and Pompey, whose tax base halved between 1986 and 1991. Custines now has fiscal resources superior to those when the iron and steel industry was active, and has welcomed most of the new firms. Without co-operation, these municipalities do not have sufficient funds to carry out useful investments. They are dependent on public funds for each of their projects. Another problem linked

Table 6.9 The level of land prices in industrial zones in Lorraine and elsewhere.

	Price per m^2		Price per m^2
Agglomeration of Nancy		*Zones undergoing conversion*	
Maxeville St-Jacques	200	STEEL AND IRON AREA	
Technopôle Brabois	115	Thionville	70–110
Fleville-Sud	85–118	Briey	42–55
Heillecourt-Est	45–92	European Pole of Longwy	50
		Villers-la-Montange	46
Pont-A-Mousson	100–120	COAL AREA	
		Creutzwald	35–61
		Valmont	20–40
		Faulquemont	10
		Bouzonville	6
Agglomeration of Metz		*Other regions*	
Marly	170	Lyon	400–700
Moulins	150	Montpellier	250–300
		Nice	250
		Grenoble	200
		Rennes	200

with the lack of co-operation is that there is no redistribution of fiscal resources at an intercommunal level. For this reason, municipalities compete with each other to attract new firms. This precludes an organized strategy for economic development.

At the moment, the private sector is absent from Pompey for both housing and industrial-estate development since the market does not offer good levels of profitability to investors. Finally, the government is intervening by means of various funds to assist investors, infrastructure spending and derelict land reclamation. Since 1984, different plans have been proposed by the state to mitigate the effects of the steel industry crisis. In addition a special prefect was appointed from 1984 to 1988 to implement these actions. Since the prefect, Monsieur Chereque, a local Pompey man, became land planning minister from 1988 to May 1991, he paid particular attention to this gloomy situation. At present, when the situation seems to have become less critical, it is intended that the level of public support should be decreased after the end of the current 1989–93 plan. This could also be true of assistance from European Community structural funds through ERDF/FEDER. The afore-mentioned EPML is another public tool charged with recycling and developing the site.

In this context, national government seems to have replaced previous actors (the local authorities and the steel company) and to have taken charge of all the economic, urban, and social development. Decentralization, which has given new responsibilities to the municipalities in land planning and development (see Chs 1.1 and 2), has not produced any positive effects here because of the municipalities' financial inability to assume the new prerogative. One of the goals of the recycling scheme is to allow local authorities gradually to become independent of public assistance. This complies with the new framework of powers.

The development of a new employment zone

The reconversion of the area has been through two phases since 1984. In 1984, the government's Steel Plan announced the closure of non-profitable steel plants, which were publicly owned. This was seen as the beginning of the end for Pompey, and several long strikes followed. In response, the government had to propose an alternative for the area that offered the prospect of rapid benefits. This explains why the state decided to facilitate the development of an industrial zone on the site. First, 40 ha of land that had not been too badly polluted was released by the steel company. New firms were given financial incentives by the state and in addition funds for the development of the zone were created, amounting to a subsidy of about FF20 per m². The regional council also contributed a subsidy of FF10 per m² for infrastructure. A publicly owned entity was created, called *Pompey Développement*, which subsidized the interest on loans and gave a bonus for each steelworker engaged in new activities.

The public steel company sold the land for a nominal price of one franc to a semi-public company, SOLOREM, based in Nancy. This organization was responsible for a ZAC that had been created on behalf of the *Syndicat mixte des zones industrielles de Meurthe-et-Moselle*. The rather pragmatic approach of re-using former roads and old industrial buildings with valuable incentives was rather successful. The arrival of important foreign firms followed, but this also created some problems. The political necessity to create new jobs, replace the disappearing steel employment, and keep to a strategy of low land prices (FF55 per m² for land with infrastructure, the average being about FF80–120 per m²) meant that firms were not selected methodically. Some of the new firms were not reliable; they purchased shabby premises, which still remain shabby, giving an image of decay to the whole area that is difficult to eradicate.

This first operation is now almost finished. About 30 ha has been sold within seven years to firms that have created 1,000 industrial jobs, but only 100 steel-workers have been employed. It should be added that this zone is essentially within the territory of Custines. Implementation on sites in Pompey is taking place last, and has not yet produced tax benefits because of a five-year tax holiday on business taxes, designed to attract new firms (see Ch. 3.3).

The second phase is in progress on 120 ha of derelict land from all over the site, which is composed of three development platforms separated by two rivers, the Moselle and the Meurthe, and a canal. The connections between these platforms and the outside are difficult because such bridges as exist allow only one-way circulation. The first platform (30 ha in Pompey) originally housed the blast furnaces, steel mills and foundries. The second platform (40 ha, the main part in Frouard, the rest in Custines) accommodated the rolling-mills and the waste heaps of a century of industrial activity. The third (10 ha in Frouard) was a metallurgical site. The remainder of the site was waste ground liable to flooding.

The first attempt to develop these three zones continued the method chosen in 1984: to attract new firms by presenting land and old buildings at low prices without a development plan. But by 1987, this method had shown its limitations and was deemed to have failed. The location of marginal firms or the conversion of old structures was perceived to make these zones unattractive. At the same time, a regional policy for derelict brown land was being created in Lorraine. In 1987, the government and the regional council decided to entrust the EPML with the treatment of the 120 ha of brown land. The approach of the EPML is based on two points: global public ownership of the site in order to control its evolution, and recycling and greening of the land before any new use.

The first objective was achieved in 1988. It was difficult to assess land prices and they have been determined by specialized land surveyors from Lorraine's land taxation centre, integrating data from the local land market. Technical information about the constraints presented by the nature of the sites was also recorded. The steelworks has sold 120 ha of land for FF7 million, together with buildings of about 50,000 m^2 for FF5 million, and other land for FF2 million (at about FF1.7 per m^2, which is the price for agricultural land).

Land recycling is now almost finished. After several technical studies, which showed the difficulty of building on sites with solid foundations, a decision has been taken to move 300,000 m^3 of soil from one site to another, to create flat platforms above the level liable to flooding. This will enable EPML to build new structures and infrastructure.

This work has enabled EPML to prepare 40 ha for a new development, with a budget of FF15 million (FF37.5 per m^2). Soil contamination has also been studied, especially on the slag-heap site, which was most affected. Site recycling also includes landscaping, which is very important for this valley site as it is visible from neighbouring towns, the highway and the railway line. One of the aims of the project is to develop a green landscape that will change the image of the site and prepare the axis of development for the new zone. Costs of this work were calculated at FF8 million (about FF10 per m^2

for 80 ha). The first part of this project was completed in spring 1991.

Lastly, some buildings were refurbished, when new occupants were found, in 1990. The EPML also evicted some fringe firms installed in old buildings, and demolished every building unsuitable for reconversion, despite the municipalities' desire to keep them and to welcome all activities, even those that cause contamination.

Within the framework of derelict land policy, the rôle of the EPML is to arrive at an end state. It offers reclaimed land to the local authorities or to investors for development. In the case of Pompey, however, the local authorities concerned could not assume responsibility for the large investments necessary to develop the zone, although they were willing to work together for a common goal.

One solution might have been to leave the site empty, as a land reserve in the northern part of Nancy. But this approach, which has been taken on other sites in Lorraine, was not applicable to the local situation for two reasons. First, if the municipalities do not have the means to realize an important project, they can develop low-quality zones that will attract undemanding firms looking for large, cheap plots. In fact, the determination of Pompey and Frouard, the two municipalities most affected by the crisis, to welcome new firms is so strong that they insist on getting the land from the EPML and encouraging interest in the large sites and low prices. It is not politically possible for the EPML and the state to block such an approach, even if it appears to waste all the effort involved in achieving real revitalization. The government must propose alternatives. Secondly, not only does this site offer some economic development on an important zone of formerly derelict land, it also balances the existing industrial development of Nancy, which has taken place more to the west and south, and thereby strengthens the region's economy, especially on the metropolitan axis Nancy–Metz.

Central government, and especially the urban affairs ministry, considered that site development was a regional challenge justifying exceptional effort. In May 1991, therefore, the government decided to extend its involvement by completing the development of the zone with the EPML as a technical adviser, and authorized the monitoring of this operation by an inter-ministerial decision-making body.

An impact assessment had been undertaken before this decision, in 1990. Using national funds, this assessed the likely success of the zone in the context of severe competition between industrial areas (Auguste-Thouard 1990). The study showed that the problem of employment zones in the region is not one of quantity but of quality. There is a lack of zones situated near the Nancy–Metz highway, visible to travellers. Firms also encounter difficulties in finding large plots of land (larger than 5 ha) in existing zones. The pattern of consumption of industrial land in the agglomeration of Nancy (20 ha per year) demonstrates that it is possible to implement an important industrial zone on the former steel site under the following conditions:

○ establishment of links between the highway and the zone by building bridges above the Meurthe and the Moselle rivers
○ phasing in of road links, beginning with an intersection to the highway
○ definition of responsibilities for the project and marketing, and
○ determination of land prices at a level of about FF100 to FF150 per m^2 in regional zones of quality.

This project was discussed for six months and accepted by local authorities. The EPML's overall responsibility for the operation was also accepted, even if local authorities initially wished to have control of their own zones in their local territories. The EPML subsequently studied the planned cost of the project. The plan envisages creation of a link between the Nancy–Metz highway and the town of Pompey, which will cross the River Meurthe to the first platform, the River Moselle to the second platform, and the railway line to reach the town. This east–west axis will link both the industrial zone and the urban areas, which may themselves be enhanced by it.

Within the zone, 43 ha is devoted to industrial activities, 13 ha to services and research, and 4 ha as a reserve for housing close to the town. It is assumed that 50 jobs per hectare should be created in the industrial sector and 100 jobs per hectare in the tertiary sector. On this basis, 3,500 jobs should be created on the completed zone. If the 3,500 new jobs are added to the 1,000 existing jobs in the first zone, the previous level of employment on the site would be arrived at.

The total cost of the operation will be about FF150 million for 60 ha, which will sell at FF250 per m^2. The costs break down as follows:

○ FF2 million for the land without buildings
○ FF24 million for land recycling
○ FF61 million for the main infrastructure, the intersection with the highway and the bridges
○ FF43 million for other infrastructure, and
○ FF20 million for financial charges, fees, etc.

Projected revenues consist of FF74 million for plot sales at FF125 per m^2, and FF77 million from subsidies (FF41 million was from the state, FF12 million from the regional council, FF10 million from the departmental council, FF7 million from the European Community, FF5 million from the local authorities and FF2 million from the EPML). This balance sheet shows clearly that the regional land market could not fund such an operation without massive public aid, including the cost of reclamation and infrastructure.

It should be emphasized that at FF26 million for 80 ha (FF32.5 per m^2), the cost of the reclaimed land is more than twice that of alternative land for future industrial use in the same regional location (FF15 per m^2). The need for public intervention in order to recycle such sites is obvious. It is important also to note that the major infrastructure, at FF61 million, is vital as it is only through the provision of good links that land prices comparable with those elsewhere in the region can be achieved. Otherwise, the selling price of the land would have been fixed at FF60–80 per m^2, which would have involved a loss in revenues of about FF26 million to FF38 million.

At the moment, the operation has just begun, so it is still too soon to draw conclusions on the economic effects of devoting considerable public funds to this project. But it is already possible to attempt an analysis of this operation, as a model for a development project on a derelict site in a situation of industrial decline.

The effect of the reconversion on the tools and the actors in land planning
Pompey is a good example of the challenge that local and public authorities face in reviving old industrial regions. A general point is that physical planning does not play a positive rôle in the reconversion of brownfield land. In fact, the first structure plan, that for the Lorraine metropolitan area, set out in 1975 (and the local plans of the municipalities defining land-uses), have only recognized the existence of the steel plant in designating the bottom of the valley for industrial activities. For this reason, the decision to develop a new employment zone on the former industrial site does not involve any alterations to the plans. In this case, plans have a conservative rôle. Because planning tools have been conceived in a context of growth, they do not help in recycling such sites. In other words they concentrate on allocating the uses of agricultural and natural lands. Moreover, current plans can constrain the redevelopment of derelict land because agricultural and natural areas are protected.

On previous steel-making or coal-mining sites in Lorraine, when new uses for leisure, housing, etc., have been decided upon, problems occur because it is difficult to change a master plan that affects several municipalities. One result of decentralization is that modification of master plans has become a complex process involving decisions from all the municipalities concerned. For example, in the case of an important leisure project supported by the public authorities on a derelict site near Metz, the administration succeeded in demonstrating that a leisure park was a particular kind of industry. There was, therefore, no need to change the master plan to realize the project. Other less important projects must wait for changes in the prescription of the master plan, or be carried out within the framework of local plans that are compatible with these master plans. These situations create conflicts, but the urgency of redevelopment sometimes requires a reinterpretation of adminis-trative rules.

The second point that must be stressed is the difficulty of achieving decentralization. Culturally, local authorities were not prepared to assume this new rôle because they were accustomed to the ascendancy of the steel firm, which provided wealth, and organized social and political life. Moreover, decentralization took place in 1982–3, during the process of decline and the closure of the Pompey plant. It is obvious that financial redistribution did not follow the trend towards decentralization. It was particularly difficult to cope in the face of the economic obstacles caused by the decline of the old industry, as local authorities hitherto dependent on it were facing huge financial problems with the decrease of business and land-related taxes.

It is clear that local authorities have not played any significant rôle in the reconversion of the steel site. The first development zone was undertaken at departmental level. The municipalities are now beginning to undertake small operations outside the perimeter, but they are not ready to complete larger projects that match the scale of the problem. The Custines project, development of 20 ha for industrial activities on agricultural lands, shows how a global approach is difficult to implement.

The weakness of local authorities partly explains the important rôle of the state, which immediately took charge of the crisis through its national steel plan. The government had to be particularly resourceful if it was to attract new firms to the site. Early experience nevertheless showed that successful reconversion needed a strategy of development as well as incentives for new industries. The inability of local authorities to establish and carry through such a strategy has forced the government to use a public organization, the EPML, to achieve this purpose.

This evolution is reminiscent of the British experience of urban development corporations (UDCs), set up by the Conservative government in the 1980s to reconvert derelict urban areas. A comparison of the main differences between these systems may help to convey a better understanding of the method used in Lorraine.

The first point is that the UDCs were conceived in order to follow the government's strategies, in opposition to local authorities' policies. Power over planning was completely transferred to the UDC. In Pompey, local authorities maintain their responsibilities, and the recycling project is discussed with them. In addition, the public subsidies are brought into play by several levels of organization (local, departmental, regional, national and European Community).

The second point concerns the economic approach. The British approach is based upon the principle of a complete return of public funds to the government through the sale of plots, which may have been understandable in London but is less so in, say, Liverpool or Manchester. In Lorraine, and more precisely in Pompey, we have discovered that the level of the land market is so low that strong public support is essential to the reclamation and redevelopment of derelict areas.

It would be possible to invest less, and to sell the site with minimal reclamation in order to limit inputs from the public sector, but this would create a marginal zone with few, unskilled jobs. In this case it would be very visible from the main axis of the region. It is not possible to undertake this kind of reconversion project if it is merely based on short-term financial returns. The investment effects of recycling, or of creation of new buildings in a valley concern not only the balance of the development zone but also its capacity to become an active part of the agglomeration of Nancy once again.

The third point, which is linked to the second, concerns the conception of the project. A profitable approach implies the rejection of every expense that does not have identified returns. The experience in Lorraine shows that derelict areas needed prior upgrading of land before any new use was possible. This is not only for technical reasons but also to reverse the bad image which is a fundamental obstacle to every new investor.

The initial effort is thus necessary, even if it offers no prospect of economic return, because it allows the improvement of the region's environment. It is important to distinguish this phase, which requires public action, from the phase of development, which must integrate with the market context.

Table 6.10 Timetable for the project.

1984:	National Steel Plan; closure of some parts of the Pompey steel plant
1985:	Creation of the development zones of Pré-à-Varois and Pompey industry (40 ha). Arrival of first new firms (Clarion, Sofreb)
March 1986:	Beginning of the regional policy for derelict lands (associating the government, the regional council and the EPML).
September 86:	Decision of the regional board for derelict land to take into account the site of Pompey
December 86:	Final closure of the Pompey steel plant
1987:	Carrying out of technical studies on site by EPML
December 87:	Signature of an agreement between EPML and local authorities.
August 1988:	Purchase of the area by the EPML (120 ha).
1989–91:	Treatment of the site (creation of platforms, demolitions, refurbishment of old buildings, landscaping) by EPML
June 1990:	Completion of an economic study about the potential of the zone (paid with government funds)
December 90:	Agreement on a global project for the site by the local authorities
May 1991:	The first development zones (Pré-à-Varois and Pompey Industries) are almost completely developed and sold. Decision of the government to support the global project and to authorize EPML to conduct the operation

To conclude, time is essential in this type of process. On the first reconversion zone of 40 ha, the activity had ceased in 1984 and the development area was created in 1985. After six years, the zone is almost completely sold. On the main site of 120 ha, the activity ceased in 1986. Five years later, the land was reclaimed and a development project was ready for implementation. The aim is to sell the first plot in 1993, and to complete the operation by 2003. The steps in reconversion are complicated, because of technical, financial and political difficulties. It is not argued that any one specific form of organization should be proposed, but public intervention is clearly necessary in order to ensure that urban and economic life is able to start again at a competitive level.

6.3 Two cases of industrial settlement in Alsace

The local and regional context

The region: competition for space Alsace is characterized by a rather high density of development. In spite of its rural landscape of beautiful scenery, vineyards, mountains and medieval castles (as will be seen, this is one of the development factors), industrial employment in Alsace is higher than the French average: 35% of the EAP, compared with a national average of 26%. The relatively high density of highways, and the presence of viniculture, rich agriculture, high-density forested areas, and very diffuse urbanization lead to strong land-market pressure. In the 1970s, one of the highest rates of growth of land prices was in Alsace.

Economic planning Except in some small mountain areas, Alsace has not benefited from state development grants such as the Prime d'Aménagement du Territoire since its level of economic development is higher than the average. For the past three years, Alsace has been the tenth most industrialized of French regions. Nevertheless, local authorities have stressed the economic disparity within the upper Rhine Valley between Baden-Würtemberg in Germany, the Basle region of Switzerland, and Alsace, as a result of which this area eventually received some grants. Before the 1982 decentralization, the Alsace Region prepared an economic development plan, the Schéma d'Orientation et d'Aménagement de l'Alsace, which was approved in 1976. This document concentrated on economic development, but covered other matters including an important chapter on the environment. This plan was not legally binding, but there is no doubt that it has enhanced the action of the institutions and administrations.

Since 1989, attempts have been made to update this plan, with emphasis on Alsace's evolution as a frontier region in the context of the Single European

Market. But even if this document includes most of the regional actors, its concrete impact will suffer from the legal impossibility of imposing measures on the other authorities, the department and districts. In spite of decentralization, in terms of planning the only (and very rarely used) means of imposing measures on lower-level local authorities are in the hands of the state.

An important contribution to economic planning (although it is far from being spatially important) is the Contrat de Plan Etat-Region. Established for a five-year planning period, this contract defines the policies and actions, which will in turn be financed accordingly. Each planning period has its priorities (the recent tenth plan stressed roads and training for workers), but there is a high degree of flexibility. In the Alsace region research and technology industries were considered priorities.

In spite of the legal principles, it is at the departmental level that the most important work is done regarding economic planning. In Alsace, the southern department (Haut-Rhin) has an especially active development policy. It attempts to attract innovative high-tech firms following the technology park concept. This has included sending an envoy to Japan whose task is to build up a network of relationships with Japanese firms and institutions. Before this was done, a survey of the most attractive settlement areas was established, and at the same time some 600 ha that had been formerly planned for industrial development were re-zoned for other purposes.

Spatial planning In the 1970s, the whole department was covered by master plans (Schémas Directeurs) founded on the rather unrealistic forecasts of growth rates characteristic of that time that are now completely obsolete. After a period of general decline in physical planning, its necessity is now increasingly felt, especially at the regional and departmental levels. A global plan for Haut-Rhin is presently being prepared, but is not yet complete. It will provide a framework for a revision of the master structure plans completed by recommendations for local planning. But again, it is not within the department's legal jurisdiction to impose planning measures at the lower levels, i.e. on the municipalities.

As explained in Ch. 3.1, the only legally binding physical plan remains the local plan (POS). In 1987, 68% of the Haut-Rhin department was covered by approved local plans. This is above the national average of only about one-third of the territory, affecting 31% of municipalities. A large number of these plans are in the process of being revised.

The relationship between economic and physical planning
It is clear that industrial settlements do not follow the same logic as physical planning. A specific development board, the Comité d'Action du Haut-Rhin (CAHR – Upper Rhine development organization), advises firms interested in

locating in industrial areas. One of the two cases studied here illustrates how the local spatial plan is adapted after the economic decision to locate a firm.

This procedure seems easy to accomplish, since the French planning system does not have strong coherence at the different spatial levels. The coherence of a local plan with the master plan is indeed prescribed, but it can be flexible in two ways: there is a provision that allows the prefect to decide the revision of the Schéma Directeur whenever an approved local plan includes some contradictory provision; and the local plan (POS) can be revised by the municipality, anticipating the lengthy process of revision of the master plan and rendering a large number of master plans in effect obsolete.

The two projects: Sony and Ricoh

The case study concerns the location of two Japanese firms in Haut-Rhin. The first is a Sony production unit employing 1,500, located in Ribeauville. Since November 1986 it has been producing compact-disc players and electronic parts for video cassette players. This factory is not Sony's first in France and produces for the whole European market. Alsace's proximity to the German market (the most important European consumer market in electronics) and intensifying linkages to that market were the main reasons for its choice as a production location.

The second Japanese project, Ricoh's plant of 200 employees in Wettolsheim (a few kilometres south of Colmar), has been running since May 1988. The group's first plant in France, it manufactures photocopiers and assembles fax machines.

Sites Sony chose a 28 ha plot at the intersection of two departmental roads, in the area of the three municipalities of Bergheim, Ribeauville and Guemar. It was classified as an industrial zone in the local plan. Ricoh also set up its plant on unserviced but agriculturally rich grounds. The site, called Erlen, covers 21 ha along a railway line and is next to the north–south main road. About 10% of it was municipal property and the rest belonged to 38 different land-owners. In this case, the local plan had to be revised, since Ricoh did not wish to locate in a typical industrial zoning area such as the one situated between Wettolsheim and Colmar.

The quality of the landscape was important in both cases: neither firm wanted to settle in ordinary industrial sites but in quiet rural areas. The presence of vineyards around (and even on) the plot in Bergheim was a strong argument for Sony; the view of the "Three Castles" similarly influenced Ricoh. Moreover, in order to enjoy this view, Ricoh purchased the plot on the mountainside adjacent to the railway, instead of an equivalent site in the plain with a less attractive landscape.

Local plans Sony's site was zoned as an industrial area in the local plan. The fact that Ricoh demanded a non-industrial area necessitated revision of the local plan for Wettolsheim. This was achieved within the very short period of less than three months. Formerly, the site of Ricoh was defined as NC (an agricultural area) in the local plan. Furthermore, it had been classified as an archaeological site, in a rather ecologically sensitive area with a beautiful landscape. It was therefore a natural area in the master plan (SDAU) of the area of Colmar–Sainte Marie aux Mines. Not surprisingly, local citizens opposed the revision of the master plan.

Timetables of the two projects The timetable of the Sony project is shown in Table 6.11.

The Ricoh development benefited from the Sony experience. This explains its rapid decision-making process. During 1986 the CAHR, through its Japanese representatives, showed several sites to the firm. It was then clear that the Wettolsheim site was the most convenient. The municipalities concerned, Wettolsheim and Colmar, undertook the first steps even though Ricoh had not announced its final choice.

Financing Sony's real-estate investment was FF36 million, which was realized by Alsabail (which has leased the land to Sony). Of this, FF18 million was advanced as an interest-free loan by the region and the department, the region's share being FF3.6 million. The state granted FF4.5 million through DATAR). Total investment was FF100 million. The access road was completed by the inter-municipal association.

Ricoh's land-purchase and infrastructure costs came to FF7.5 million, advanced 50:50 by Colmar and Wettolsheim municipalities.The price of the plot was about FF2 million. The department gave a grant of FF200,000.

Land purchase and land prices Land acquisition for the Sony plant was carried out between June and December 1985. There were 35 owners on the 28 ha allocated for Sony and most sold without much disagreement. A public meeting provided land-owners with information on the project, and announced attractive purchase prices. Although the land was valued at FF6.50 per m², the inter-communal association offered FF 12.50. Acquisition was therefore not particularly difficult. Some land-owners did, nevertheless, refuse to sell and were eventually expropriated. This delayed the operation by six months.

Ricoh's land-purchase process was similar to Sony's: 38 owners shared the designated area of 22 ha, about 10 per cent of which was municipal property. The land was valued at FF6.50 per m², but the price offered was FF15.00. The only expropriation carried out was on 11 ha, and the whole acquisition took place within three months.

116

Table 6.11 Timetable for the project.

1983:	The CAHR gives lectures at the Chamber of Commerce and in Tokyo (KEIDANREN) and gets in touch with Sony's representatives
6 Jan 1984:	The president of Sony France meets the representative of the CAHR in Tokyo
18 Jan 1984:	Various sites are proposed to Sony in Upper and Lower Rhine
20 Feb 1984:	Visit of Sony representatives to Alsace
	Grants are offered to Sony for location in Alsace
Spring 84:	Sony shows interest in a location in Alsace, but there are other sites in competition, in particular in Wales and Germany
Autumn 1984:	Sony confirms its interest in investing in France, but Alsace is not the only possibility
Feb 1985:	A Japanese delegation visits 13 sites and nine firms in Alsace
5 March 1985:	Meeting at DATAR to discuss the possibility of an exceptional grant
19 June 1985:	Sony announces its location choice in the area of Bergheim–Ribeauville–Guemar; an association of communes is created to acquire the land from the farmers on a voluntary basis, i.e. without expropriation
31 Jan 1986:	The planning permit is granted by the mayor of Bergheim (less then three weeks after application); Sony organizes an architectural competition
Feb 1986:	The intercommunal association sells the land for a symbolic FF1 to the firm Alsabail (a financial agency of the regional council), which in turn leases the land to Sony for a 15-year term; furthermore, the intercommunal association undertakes the works for the access road
1 March 1986:	The building of the factory begins
9 April 1987:	Inauguration of the Sony plant
April 1990:	Completion of a second production unit; construction of the third production unit which became operational at the end of 1990 (between 1986 and 1990, the surface area of the plant increased from 9,800 m^2 to 38,300 m^2 and the number of employees from about 200 to 1,500

Actors The Japanese firms did not know France well, so needed good contacts with local institutions in order to appraise large amounts of information about local behaviour patterns and the intricacies of the local decision-making process on which to base their own investment decisions. The Japanese firms' decisions to site themselves in Alsace were based on very

Table 6.12 Timetable for project.

9 Jan 1987:	The municipality of Wettolsheim decides on the revision of the local plan and asks for an association framework with the municipality of Colmar
9 Feb 1987:	The statutes of the association are approved with the task of purchasing and servicing the area; the project is approved and the local plan is revised
3 March 1987:	Dissenting opinions from the environment ministry; however, this does not affect the revision procedure (see Ch. 3.1).
March 1987:	Ricoh decides to locate in Wettolsheim
Spring 1987:	Land is purchased
Nov 1987:	The foundation stone is laid
4 Nov 1987:	Ricoh applies for building permit
30 Jan 1988:	A building permit is granted by Wettolsheim s mayor
May 1988:	Beginning of production (with 80 employees)
October 1988:	Inauguration of the plant (200 employees); since then, Ricoh has built an extension of the plant (440 employees in June 1991); the second extension should be complete in March 1992, employing another 160 workers

precise and explicit criteria. Among these were the attitude of workers and the quality of landscape around the sites. But the internal decision-making process of the firm involved many people and cannot be explained in detail in this case study.

In both cases, the CAHR monitored the whole process of industrial location, from the first contact to the start of production. It selected various sites, organized the visits for firms' delegations, collected statistical data, contacted the state and the municipal administrations, and intervened on behalf of the firms.

The Colmar chamber of commerce and industry conducted the land acquisition on behalf of the municipalities. It organized public meetings, fixed the price offered to private owners, contacted every land-owner concerned to negotiate contracts, and initiated expropriation where this proved necessary.

By giving an exceptional grant, the DATAR office facilitated the establishment of the Sony plant. Indeed, Sony would not have agreed to come to Alsace without it. Ricoh did not obtain such a grant. There had formerly been a conflict between DATAR and the CAHR, which is the only region to have a delegation in Japan, as DATAR had some reservations about the wisdom of attracting Japanese plants. Furthermore, DATAR had favoured other, less-developed, regions when they had negotiated with the two firms.

The DDE of the urban affairs ministry at the departmental level revised the local plan (POS) of Wettolsheim on behalf of the municipality. It also

118

consulted the other land-use related administrations as prescribed in the procedure. Although it faced intense opposition from the culture and environment ministries (see Ch. 3.1), the DDE office nevertheless strongly defended the local plan's revision.

The mayors of the municipalities involved were also wholly in favour of the two projects once they understood their industrial consequences. Although it is usually difficult to obtain cooperation between municipalities, this did not cause any problem in these plant locations. In both cases, an association (*Syndicat Intercommunal*) was created, and the costs of attracting these firms were shared. The demands of the firms (like Ricoh's desire for a precise site location of its choice) were satisfied and the building permits granted within very short time limits.

Conclusion

The Sony and Ricoh case studies demonstrate the favourable institutional climate prevailing during the firms' location process. State administrations, mayors, chambers of commerce and other economic organizations all cooperated efficiently to make the developments possible. The initiative came in these cases from the organization of the department, the CAHR. The municipalities were informed only when the firms showed a serious interest.

It is clear that physical planning plays a rather marginal rôle in the decision-making process for industrial location. The area could have been zoned as an industrial area or as a non-industrial area, depending on the arguments for or against both firms' location. Planning has therefore been adapted *a posteriori*. This can be explained by the very weak global conception of planning at the regional level.

A particular aspect highlighted by these cases is the interest expressed by both firms in the quality of landscape. In coming years, this could become one of the major characteristics of this location for high-technology or service firms that wish to improve their image through the quality of the environment. An obvious paradox appears: the eventual damage to an otherwise outstanding landscape could diminish the attractiveness of the region, in particular to the tourism sector. For the time being, however, the institutions are mostly orientated to the short term, and the scenery aspect has not yet strongly affected physical planning. However, discussions and work on comprehensive planning at the departmental level shows an increased tendency for landscape concerns in location matters.

PART III
The property market

CHAPTER 7

The framework within which
the urban property market functions

7.1 The legal environment

Law, acts, powers and plans

In this field, the basic text is the "general rule of floor area", which deals with location, structure and orientation of buildings. The local plan is a public document available in town halls and in the departmental divisions of the urban affairs ministry (see Ch. 3.1). This specifies the floor-area ratio (FAR) or the building density in relation to the land area. For example, a FAR of 1.5 means that a surface of 3,000 m² can built on a land area of 2,000 m². If this ratio is not respected, a tax can be levied.

There is also a legal density ceiling ratio (LDC) that fixes the ratio between the building surface (without inside installations) and the land area. The LDC is fixed at 1 for the whole national territory, except for Paris, where it is 1.5. A tax is also collected in case of excess. The FAR and the LDC may both apply. The goal is to avoid overdevelopment of the urban environment. A planning certificate giving information on land-use conditions can also be obtained at the town hall. This determines whether the land can be built upon or not in relation to the existing urban context, public utility services, and any rules defined in the plan unit development, subdivisions or local plan. It defines the urban rules to be respected, such as the building density, the size, and the technical, financial and administrative conditions (drainage, servicing of land, local tax, building licence, etc.). The building rules are precisely defined by certain specific codes, for example, the building and dwelling code for housing. The aim of these regulations is to provide security, salubrity and environmental conservation.

New buildings

As far as the technical aspects of building are concerned, the general design can be implemented by anyone with the required qualifications, for example, an architect or building foreman, or by a company that employs such people. The right to build is one of the fundamental elements of the property right defined by common law. But to benefit from this right, it is necessary to respect the prevailing regulations.

The building permit is the administrative agreement that establishes the conformity of the project with land-use regulations. The details of the building permit procedure are specified by the urban code to give the maximum guarantee to the users. The building permit must be acquired from the municipality. When it is delivered, it must be made public. If the proposed work respects the building permit and the prevailing regulations, the manager of the departmental division of the urban affairs ministry delivers a "certificate of urbanism".

A demolition permit must be obtained before a building is wholly or partly demolished. The request is submitted by the owner and documents concerning the goal of the project must be enclosed. The procedure is then the same as that of the building permit.

Old buildings

An old building can be subject to easements. When an old building is to be bought, a planning certificate, as in the case of new buildings, may be required. This request is not necessary but it constitutes an important security for the buyer as it reveals the history of the building.

Improvements on old buildings Depending on the type of improvements to be carried out, it may not be necessary to employ an architect. For improvements that do not increase the surface area of the building, the owner must only declare the modifications to the municipal services. But a building permit will be necessary if the owner wishes to change the use of the accommodation, to transform the outside aspect and the size, or to create additional floors. When the building is close to an historic monument (less than 500 m away) or on a listed site, an additional agreement must be delivered by the service for historical buildings (*Bâtiments de France*, see Ch. 3.1).

Other regulations

For all types of buildings, including private housing, there is a legal system regulating joint ownership. The joint owners must respect the obligations determining the use and the management of the building. Each joint owner bears a proportion of the costs involved for the collective services and facilities according to his or her shares in the condominium.

Public intervention

Public intervention in the building market is defined in the master plan. As already stated, it is an urban document that determines long-term planning policy. It concerns several communities and is the framework within which the local plan operates (see Ch. 3.1). The main goals of this master plan are to locate areas to be preserved from urbanization (e.g. forests, etc.), to locate areas to be urbanized, and to locate the public services and facilities necessary for present and future urban areas.

Information systems

Different information systems are available for finding real-estate property: estate agencies, which act as intermediaries between buyers and sellers; estate brokers, which buy and sell real-estate property for their own accounts; specialized newspapers and magazines; and public notaries.

These agents may give legal information but it is also possible to employ a legal adviser. The cadastral survey, which gives information about the surface area covered, land-use, taxation value, identity of the owner, etc., is open to the public in each town hall. It is also possible to obtain an expert opinion on the real value of the property on the market. For that purpose, an average price based on market value is calculated based upon the real-estate characteristics (quality, location, tenure), and the constraints in terms of public law (town planning, building regulation) and private law (joint ownership rules, etc.).

7.2 The financial environment

The general conditions of the economy within which financing operates have changed considerably since the early 1980s. The reforms that began in 1983 have led to almost total deregulation. The capital market is now consolidated and offers very short-term to long-term loans. After the creation of new financial instruments (e.g. financial bills that firms can issue), and new divisions (like the secondary market where the shares of small and medium-size companies are exchanged) most economic agents can lend or borrow on the cash. They can even utilize the futures market for capital.

The increase in the profit margins of companies and the increase in the capacity of companies to finance themselves, together with the deregulation of the capital market, led to financial disintermediation in the late 1980s. Firms prefer to borrow on the capital market rather than call on banks for credit. To preserve their level of activity, the financial institutions have turned to the private housing market and to this end have increased the supply of consumer credit and housing loans. The credit market has also been deregulated. Government-subsidized loans have almost totally disappeared except in the agriculture and housing sectors, the volume of credit that banks

124

can offer is no longer regulated, and the main instrument of monetary policy is the interest rate.

Today, this reform has led to fierce competition between the financial institutions to collect savings and grant loans. Households' new saving behaviour, the competition to collect savings and the monetary policy whose goal is to keep a balance of exchange between the French franc and the Deutschmark have led to a higher interest rate on deposits. The interest rates for loans are high even if the competition cuts down the profit margins.

Table 7.1 Interest rates and banking costs (%).

	Management cost [1]	Cost of banking resources [2]	Nominal break-even point [1]+[2]=[3]	Inflation rate [4]	Breal break-even point [3]-[4]
1980	4.0	9.2	13.2	13.7	(0.5)
1981	3.8	11.2	15.0	13.4	1.6
1982	3.6	11.0	14.6	11.7	2.9
1983	3.5	9.3	12.8	9.3	3.5
1984	3.3	9.5	12.8	7.5	5.3
1985	3.0	8.6	11.6	5.7	5.9
1986	2.8	7.5	10.3	2.4	7.9
1987	2.7	7.5	10.2	3.2	7.0
1988	1.9	7.4	9.3	2.8	6.5

In order to finance construction, developers generally create a *société civile immobilière* (SCI – private property company) that will be dissolved after the sale of the last unit. The shareholders of this company provide capital and if this is insufficient the developers can ask banks for credit. It is a short-term loan (two years), at a fixed interest rate with the main repayment due on the day of settlement.

To buy real-estate property (or a share of a property), an investor can obtain an ordinary mortgage loan from a bank. The conditions must be negotiated between the two parties. If the investors want to buy office or commercial space they can contract a leasing agreement/mortgage.

In fact real-estate financing is not very different from general investment financing except in the case of state-regulated housing. First, in order to finance investment within the social rental sector, the economic agents and particularly the social rental organization can call on the *Caisse des Dépôts et Consignations* (CDC) and on the *Crédit Foncier de France* (CFF) for a cheaper loan, known as the *Prêt Locatif Aide* (PLA). The interest rate on this PLA loan is very low because it is financed by a special type of deposit account (the *livret* A) by the saving banks and collected by the CDC. The interest rate on these deposits is low, currently 4.5 per cent, but the return paid to the household is not taxed. To benefit from such loans, the investor must promise to respect certain conditions, mainly to rent the dwellings to households whose incomes do not exceed a ceiling fixed by the government.

To get extra financial help investors can also call on companies that collect what is known as the employer contribution to the construction effort. This contribution is similar to a tax. All companies that employ more than 10 workers must pay the tax, presently 0.45 per cent of the total wages paid by the firm. It is collected by specialized companies that use it only to finance housing and, in particular, to offer loans to housing investors at a very low interest rate of interest (currently 3 per cent).

Households wishing to become owner-occupiers can resort to three types of credit:

○ The *Prêts Aides à l'Accession à la Propriété* (PAP), which is a government-subsidized loan that people can contract to CFF. The interest rate on this loan is the lowest that households can obtain, but to benefit from it a household's income must not exceed a fixed ceiling. The PAP can only finance the purchase of new housing, though old housing is permitted if purchase is immediately followed by renovation.

○ The *Prêts Conventionnés* (PC) is a loan that can be offered by all the banks that sign a contract with the state, according to which the interest rate cannot exceed a ceiling fixed quarterly by the CFF. The PC can finance the same investments as the PAP.

○ The ordinary mortgage loan, for which the banks fix the conditions freely.

Table 7.2 summarizes the main conditions attached to these loans. To obtain extra financial help and to reduce the cost of debt, households can obtain other types of credit, because the amount of the loan is normally insufficient for the complete financing of a housing purchase.

First, there are "social loans" granted by the family allowance funds, by the CFF for civil servants, or by companies that collect for the employer participation to the construction effort.

Secondly, households can contract a housing savings account. In this case, they bind themselves to make regular minimum deposits for four years. During that period households receive interest from the banks or from the savings banks and a subsidy from the state. The returns on these assets are not taxed. At the end of the savings period, participating households can obtain loans on very favourable terms. The repayment instalments and the repayment period are such that the interest households will have to pay is based on the received interest and subsidies. All the conditions are fixed at the beginning of the contract and the interest rate is lower than that of other loans.

Table 7.2 The main financial conditions with respect to housing loans.

Characteristics	Types of loans		
	PAP	PC	Ordinary
Credit security	mortgage	mortgage	mortgage
Interest rate	determined by the State fixed or variable	ceiling fixed by CFF fixed or variable	free fixed or variable
Current interest Rate level	8.61%	10.92	11.42
Term	20 years	15, 18 or 20 years	maximum 20 years normally 15 years
Maximum proportion	90%	90%	100% normally 60–70%

To finance investments in the private rental sector, companies can seek an ordinary mortgage loan from a bank. If the investors are households, they can use the housing savings system and a PC loan under the same conditions as those for home-owners.

7.3 The tax and subsidy environment

Taxes on the property market

Real property is taxed, and there is a distinction between recurrent taxes, and non-recurrent taxes, which are generally paid when a transaction takes place.

Recurrent taxes There are three main recurrent taxes (see Ch. 3.3). Real-estate taxation is paid by the owners of residential, commercial and industrial units to the local authorities. This tax is based on the rental value of the real estate after a rebate of 50 per cent. The real-estate rental value of industrial and commercial companies that own their buildings is determined by the taxation authority on the basis of their accounts. In other cases, the taxation authority estimates the rental value. For example, the housing rental value is appraised by comparison with a sample of housing classified into eight categories according to quality. All new buildings, including industrial and commercial property, are exempt from taxation for two years after completion. Housing financed by a PAP or a PLA is tax-free for 10 years (cf. Fig. 3.3).

Residence tax is paid to local authorities by all households for the

127

dwelling in which they live. This tax is based on the rental value of the house, estimated by the taxation authority as for the real-estate taxation, after some rebates for dependants, low incomes, etc.

Under the current income tax system, real estate is treated as an investment asset. If real estate is owned by a company, property income is counted in turnover and the property costs, including depreciation, as other expenditures. So the net property income is taxed at the taxation rate applicable to commercial and industrial profits. If the real estate is owned by a household, the principles are the same, but some property costs are fixed: For example, if a household rents a dwelling, it is taxed on the net return of housing according to the scale of taxation relating to personal income. But the net income is equal to the received rent less a 30 per cent rebate if the housing is new or 8 per cent if it is old. This is in order to take into account the property costs (depreciation, management costs, etc.), less also the real-estate taxation and possibly the interest payments for the loans contracted to finance the housing purchase or the housing renewal.

There are two other recurrent taxes. The tax for household refuse removal is based on the rental value after a rebate of 50 per cent; the same as for the real-estate tax. Factories and the buildings occupied by the civil service are not taxed. Wealth tax is levied on each person owning private assets whose value exceeds FF4.6 million. The rate of tax is graduated. Real-estate property falls into this category (without rebate) at its market value (see Ch. 3.3).

Non-recurrent taxes Investors must pay several taxes when they buy real property or when they put up a new building. If a company wants to build a factory or an office building in an already highly developed region, it must pay a location tax. In order to conclude the property registration, the purchaser must pay a "property advertising" tax (between 1 and 2 per cent of the real-estate property price; see Ch. 3.1). All transactions on new buildings are subject to value-added tax. The rate is 18.6 per cent and is applied to the gross real-estate market value. Transactions on old buildings are not subject to VAT, but the new owner must pay another tax, called the property transfer tax, whose rate is between 7 per cent and 17 per cent according to the type of building (dwelling, office, etc.) and the type of transaction. Agents who sell real-estate property must pay a tax on capital gains (the difference between the sale price and the purchase price). Some rebates are made on capital gains related to interest paid on loans, renewal expenditures and inflation. When the real-estate holding exceeds two years, a rebate is applied, graduated according to the length of time of the holding. Net capital gain is taxed according to the scale of the industrial and commercial profit tax or a household's income tax. Capital gains realized by a household when it sells its residence are not taxed if a new house is bought (see Ch. 3.3). As a whole, taxation on transfers of property can be considered heavy, complicated and a serious impediment to market fluidity. In addition the system does not have well defined objectives.

Subsidies on the property market

Companies may receive subsidies or may profit from tax deductions from the state or from the local authorities if they agree to build factories, offices, laboratories, etc., in certain special areas. The aim is to preserve or to develop the economic activity in these areas. At the moment, most of this direct or indirect subsidy comes from the regions, the counties or the municipalities. It varies considerably between local municipalities.

Overall, local communities do not intervene to a great extent in the housing sector. Generally, their subsidies come directly through land contributions to the social housing organizations. Subsidies and tax deductions are almost always the state's responsibility.

Social rental housing organizations that finance new buildings, or the renewal of old buildings with loans from the CDC at low interest rates, receive state subsidies. The subsidy connected to the PLA, if it is to finance a new building, is equal to 12.7 per cent of the investment cost (under certain conditions). The subsidy represents 20 per cent of the investment cost (up to FF70,000 per dwelling) if it is used to renew old social rental dwellings. In this case, the state subsidy can be supplemented by a regional and/or departmental subsidy equal to 10 per cent of the investment within certain fixed limits. Investors in the private rental sector can obtain a subsidy from the *Agence Nationale pour l'Amélioration de l'Habitat* (ANAH – national agency for housing renewal) to renovate housing units built before 1 September 1948. Generally this subsidy is equal to 25 per cent of the investment cost, but the rate can be changed according to the characteristics of the property.

Together with these subsidies, the social rental-housing organizations can obtain another allowance from the state and also from the regions and the counties. This is available when they want to build new dwellings in areas where the real land price is higher than the average one used to calculate the investment subsidy. Owner-occupiers of houses cannot receive subsidies except to renew the place where they usually live if their incomes exceed a certain ceiling. The subsidy can be higher than FF70,000 and is provided by the state.

Investors in the housing sector can also benefit from tax deductions. First, the social rental-housing organizations are not taxed on their gains. Secondly, investors in the private rental sector, if they are private persons, profit by a tax deduction if they buy or build a new dwelling. Their income tax is

129

reduced by 5 per cent of the investment cost for two years, up to a limit of FF30,000 each year for a married couple. So generally, households limit their investment cost to FF600,000.

Thirdly, owner-occupiers can profit by income tax relief if purchases are financed with a loan, for five years from the date of purchase. The reduction is equal to 25 per cent of the interest paid, but it cannot exceed a ceiling of FF30,000 if it is a new dwelling or FF15,000 if it is an old one or, since 1991, by 5 per cent of the investment cost for four years up to FF15,000 each year (for a married couple). Fourthly, owner-occupiers can take advantage of income tax relief for some home improvements.

CHAPTER 8

The process:
the property market

8.1 Price-setting

Until the early 1980s, research mostly dealt with price-setting in the housing market, and focused on supply. Price-setting was based on a governmental supply mechanism. Indeed, in the early 1980s the great majority of dwellings being built were financed by subsidized or regulated loans (PAP or PC), to which ceiling prices were applied. So the new housing prices were determined by the sum of the production costs plus a profit margin that varied with the economic trend (level of the developer's stocks, financial costs, etc.) and with a possible disequilibrium on the land market.

In the second-hand market, prices were determined in relation to prices for new housing, discounted to allow for two categories of depreciation: depreciation because of gradual decay, and financial depreciation to correspond to the sum total of subsidies allowed to new housing, making household solvency possible in this segment of the market. These mechanisms have disappeared, and today prices are set according to principles of price-setting conditions in a competitive market in which the demands are relatively price inelastic.

Since the early 1980s, the second-hand market has considerably increased; the number of transactions made by households is nowadays nearly twice as many as those in the new housing market. The government's housing policy is clearly characterized by a free-market approach. Now, with deregulation of the housing sector, price setting of new and old dwellings is based on the solvency of the agents. Suppliers adjust their production in terms of both quality and price. They determine the solvency conditions of the buyers (income level, family size, etc.) and afterwards estimate the amount the household can reasonably borrow and pay back. Given the amount of initial capital, it is then possible to assess the maximum price level of the dwellings.

The price-setting mechanism is closely linked to a selective filtering of customers. According to this analysis, one of the main indicators of housing market regulation is the "marginal relief"; that is to say, the difference between the new housing price and the old one. The origin of this marginal relief can be explained by the characteristics of the allowances and financing system for each market. The conditions making it possible for buyers to be solvent and the filtering process of the customers are not the same in the two markets. The housing market in Paris can be taken as an example despite its unique characteristics. The large increment in prices recently observed reflects the intrinsic characteristics of its situation (location and size of the housing stock, tensions on the rental sector, lack of urban space available, etc.), its reorganization (filtering of the households according to their solvency), and the hierarchy that has developed throughout the districts.

The housing market in Paris is based on a nucleus of districts from which price increases spread. But it is also a market where the solvency characteristics of the buyers are decisive, as shown by the progressive disappearance of marginal relief. A similar analysis in 65 neighbouring municipalities explains the main criteria that influence prices: proximity to the Parisian market, quality of living conditions, existing public transport, etc. However, alongside this open market, there are state-aided sectors of the housing market where prices are regulated under the conditions attaching to *Prêt d'Accession à la Propriété* (PAP) and *Prêt Conventionné* (PC) loans.

The PAP is a state-subsidized system of loans for home-ownership that was set up in 1977. Prices of dwellings subject to subsidy cannot exceed a ceiling amount that varies according to family income, the geographic area and the type of property. They can finance new housing construction and old housing purchase with the aim of refurbishing them.

With a PC loan the total amount cannot exceed a cost price per square metre that varies with the geographic areas. Accordingly, the official prices are currently determined in the same manner as dwellings in the open market. As with the PAP, PC loans originate from the housing financing reform of 1977, but they are not direct state-aided loans. They are issued by the banks, which have signed an agreement with the state. They also finance new housing construction and old housing purchases where the buyer intends to refurbish.

8.2 Actors and their behaviour

In order to analyze the property market, it is necessary to distinguish the areas according to the tensions that can exist. In the Paris region and some other large cities, the actors take positive action in order to make capital gains. For example, in Paris firms agree to sell office buildings that they own and occupy, and to buy or to rent others if they can obtain considerable capital gains by this means. On these markets, the main actors are the large firms, the insurance companies, some financial institutions such as the CFF and foreign economic agents (Japanese, Swedish, among others). But speculation is also sustained by the real-estate brokers who buy in anticipation of a capital gain, improve the buildings and resell them at a profit. On the other property markets the behaviour of agents and economic actors is connected with their needs and with some long-term considerations.

The first category of actors who intervene in the property market is the developers. It is necessary to distinguish between the social and non-social developers. The main social developers are the housing associations. They build and manage social rental dwellings, and build and sell housing to households that can obtain PAP loans. Since the beginning of the 1980s their power has increased. Some of them have changed into "building and planning council organizations" that can also plan a city area, and build and manage non-housing real estate. Most of them are owned by the municipalities or by the departments and act according to local policy. Others are owned by the banks, the insurance companies, and companies promoting employer participation in the construction effort.

The non-social developers are often very different. These are the people who carry out small operations each year, generally in the less-intense property markets and often only in the housing sector. During the 1980s most of them stopped their activities. Now, private developers are large companies, often subsidiaries of banks or of such vertically integrated building companies as Bouygues, Générale des Eaux, etc. Their behaviour is described below under price-fixing. Generally they try to intervene in tight markets, where they can fix a very high price and obtain a very large profit margin. First, they define the area where they wish to build, and fix the price according to the solvency of the prospective buyers. They then try to buy the land and offer the land-owner a price in line with the price of the units that they have fixed. The same process can be followed to renew an old building before selling it.

Property brokers are speculators in economic terms. They buy buildings with the expectation of capital gains, keep the property for a very short time and sell it without necessarily making any improvements to it. Property brokers are just intermediaries, and their profit comes from a capital gain. Another group of intermediaries is the real-estate agents. They act as middlemen between buyers and sellers or between owners and tenants. Their returns are fees upon transactions (see Ch. 4.2). Property markets, including the housing market, are now deregulated, so agents who cannot pay the price fixed by the market are excluded from it. This leads to great difficulties for households in finding a house, especially in tight markets such as in the Paris region.

The tensions in the housing market in large cities are increasing because more and more households want to live in the town centres. To solve the housing problem of the underprivileged households in large cities, the government has an interventionist policy. In the Paris region, for example, it fixes each year a maximum rate of rent increase (see Ch. 4). In order to encourage the building of social dwellings in the large cities, where land prices are high, the 1991 *Loi d'Orientation pour la Ville* (LOV – land law) was introduced. The government was then empowered to bring in a new tax that developers must pay when they start a new development. Consequently, developers are required to give a section of land, equal to 10% of the land used in the building operation, to the community in which the development is situated in order to build social housing in the neighbourhood.

Table 8.1 Housing stock by tenure and type of owner (000s of accommodation units).

| Year | Total stock | Second homes | Vacant housing | Rented stock | | | | | | Owner/ owner-occupiers main | Total main |
| | | | | Public sector | | Private sector | | Free housing | Others | | |
				Social housing	Other lessors	Legal entities	Individuals				
1973	20,321	1,591	1,606	1,933	297	671	4,060	1,747	618	7,798	17,124
1975	21,079	1,686	1,607	2,188	319	652	4,062	1,698	579	8,288	17,786
1978	22,236	1,844	1,751	2,481	342	697	4,132	1,656	537	8,796	18,641
1982	23,706	2,248	1,784	2,643	355	635	3,964	1,636	513	9,929	19,674
1984	24,249	2,288	1,898	2,898	365	601	3,859	1,616	475	10,279	20,093
1986	24,723	2,358	1,968	2,850	375	571	3,794	1,595	460	10,751	20,397
1988	25,177	2,439	2,038	3,144	397	541	3,544	1,495	346	11,233	20,700

Source: M. Mouillart & Y. Martin 1989.

CHAPTER 9

The outcome of the
urban property market

9.1 Distribution of ownership and tenure

Both Tables 8.1 and 9.1 show the housing stock (by type of tenure and
typology of the owner) and the rate of ownership per sector.

Table 9.1 Trends in property ownership.

Year and source	Households owning their own residence (%)
1973 EL	45.5
1975 RGP	46.6
1982 RGP	50.5
1984 EL	51.2
1988 CEREVE	54.5
1989 OEM	54.7

Source: M. Mouillart & Y. Martin, 1989.
Key: EL – Enquete-logement (housing survey); RGP –
general population census; OEM – Observatoire de l'
Endettement des menages.

Some trends are discernible: there are more owners in spite of a slowing
down of new construction (see Ch. 1.4) and the rate of ownership was 54.3
per cent in 1988, up from 51.2 per cent in the last housing survey in 1984.
Home-ownership expansion continues at the same rate as observed previously because of growth in the second-hand market, where an increase in
transactions between 1984 and 1988 of more than 40 per cent was observed.
By comparison, sales of new dwellings were going down during that period.
The collective sector is also important in the trends in transactions because
of the new attraction of the town centres (Table 9.2).

Table 9.2 Second-hand market development (000s).

Year	Rented	Sector Home owners	Second home	Total
1980	46.7	261.4	43.9	352.0
1981	49.4	248.0	47.6	345.0
1982	45.5	294.8	32.1	312.4
1983	38.0	290.0	31.5	359.5
1984	14.7	330.5	26.5	371.7
1985	7.8	337.3	26.8	371.9
1986	12.7	352.9	30.8	396.4
1987	34.9	393.2	31.2	459.3
1988	50.7	413.1	35.8	499.6
1990	45.1	420.9	33.7	499.7

Source: B. Lefebvre, M. Mouillart, S. Occhipinti, "Perspectives du secteur et du financement du logement à l'horizon 1993", from a conference at the Assemblée Nationale, 21 June 1990.

Changes in distribution

Recent home-owners, who bought their dwellings between 1 January 1985 and the time of the housing survey in 1988, were mostly first-time buyers. Those who were already owners were more numerous in 1988 than in 1984, a fact confirmed by the slight development of sales in cash. This market is fed by mobile homeowners who resell their previous dwellings, for example, inherited properties. On the other hand, the proportion of households who live in inherited dwellings keeps declining (7.1 per cent in 1988 against 8.7 per cent in 1984), the result of inheritance at a later age, by which time the heir's housing needs are already satisfied (Table 9.3).

In the rural communities, where nearly all dwellings are individual units, more than one-third of households have bought their own dwellings. In rural

Table 9.3 Home owners in 1984 and 1988 by origin of the property.

	1984	1988
By inheritance or gift	8.7	7.1
Cash purchase	11.8	14.0
Including: resales of previous house	(52%)	(54%)
Grants and subsidies	(8%)	(6%)
Credit purchases	78.1	77.9
Including: resales of previous house	(21%)	(22%)
Grants and subsidies	(10%)	(9%)
Annuity life purchase	1.1	0.7
To-rent purchase	0.3	0.3
TOTAL	100.0	100.0
Total (000s of accomodation units)	1,722	1,802
All recent buyers (000s of units)	1,736	1,821

areas, and also in small towns, housing chosen from builders' catalogues predominates, with 45 per cent of the market. At the national level, the share of "catalogue" housing has increased since 1984, while the share of "beavers" (do-it-yourself) has stabilized. In the second-hand market, purchasing modes vary a lot according to the locality. Real-estate agents intervene in 40 per cent (on average) of the transactions, but their share of the market is equal to 21 per cent in rural areas, contrasting with more than 58 per cent in the Paris outskirts (see Table 9.4).

Table 9.4 Recent homeowners by type of municipality and property market (%).

	Rural communities	Urban units (population)		Paris & suburbs	Total
		< 100,000	> 100,000		
New housing					
Individual housing					
Built by the owner	10.4	6.7	3.3	6.3	7.4
Built by an entrepreneur	43.9	32.1	26.2	23.1	34.9
From a catalogue	44.5	45.7	28.5	19.1	39.6
From a developer	0.9	9.7	20.2	23.5	9.5
Flats	0.3	9.7	20.2	23.5	9.5
TOTAL	100.0	100.0	100.0	100.0	100.0
Second hand market					
bought through					
A real estate specialist	20.7	33.9	43.7	58.4	40.0
A notary	23.8	22.4	10.2	3.7	14.5
Bought directly	55.5	43.7	46.1	37.9	45.5
TOTAL	100.0	100.0	100.0	100.0	100.00

Distribution of ownership according to mode of finance

Table 9.5 displays the structure of the financing scheme in the buying of residential property. Different choices are offered to buyers, especially in the new housing sector: PAP and PC loans. The first are subsidized loans for home-ownership and the latter are state-aided loans.

The trends observed between 1984 and 1989 reflect the recent evolution in financing conditions. The decline of new housing construction is the result of the state's withdrawal from the housing sector (48.1 per cent in 1984 to 17.7 per cent in 1989 for the PAP in new housing, and 7.7 per cent to 1.4 per cent for the PAP in old housing during the same period). As for the PC, their backward movement shows the opening of the housing market to the

Table 9.5 Home ownership by financial resources (%).

	1984	1985	1986	1987	1988	1989
New housing	100	100	100	100	100	100
PAP	48.1	40.9	37.3	31.5	23.0	17.7
PC	39.2	46.2	42.8	46.0	41.5	38.7
Owner-occupiers	2.0	2.8	8.4	9.2	19.1	23.0
Rented housing	2.4	1.4	1.6	4.0	6.3	9.6
Second homes	8.3	8.7	9.9	9.3	10.1	11.0
Old housing	100	100	100	100	100	100
PAP	7.7	6.6	4.1	2.5	2.0	1.4
PC	12.0	15.2	16.2	15.4	7.6	8.1
Owner-occupiers	69.2	68.9	68.7	67.7	73.1	71.4
Rented housing	4.0	2.1	3.2	7.6	10.1	11.6
Second homes	7.1	7.2	7.8	6.8	7.2	7.5

Key: PAP – prêt d'accession à la propriété; PC – prêt conventionnés.

banking system. This type of loan is thus more prone to monetary fluctuations (see Ch. 8.1). The development of the second-hand market is the result of the private lessors' withdrawal, but is more and more the consequence of important financing methods used by the banks which are seeking to gain a foothold in the market. Households become home-owners in the open sector (or free sector) with second-hand housing (and generally without improvements).

Benefits of owner-occupation

The French taxation system grants different advantages to owners according to the type of deal, for purchasing a main residence for owner-occupiers, and for purchasing housing with the intention of renting it for lessors. There are three main advantages offered: tax deduction in respect of loans, calculated on the interest paid for the amount borrowed; exemption from real-estate taxation or building property during the year of the construction and for two years afterwards; and exemption from capital gains taxation. In cases where accommodation is bought for the purpose of renting, a flat tax deduction is applied in respect of the first 25 per cent of the rent collected. This applies to the first 10 years the property is rented.

9.2 Prices, rents and speculation

Current trends for prices in the urban property market are different in the Paris region and in some other large cities compared with the rest of France.

Information to illustrate this is scant (see Chs 1.4 and 1.5). As regards rental levels, location is an important factor (see Ch. 5.3). It is necessary to distinguish between the Paris area and the rest of France. In Paris in 1990 the average housing rent increased by 7 per cent, compared with 5.2 per cent for the whole of France despite the fixed increase rate set at 3 per cent by the government for Paris for all current leases. The increase is therefore due to the rent fixed in new leases.

Speculation in property

The activity of the property broker is essentially based on speculation and so involves high risks. Speculation is usually defined as an operation that takes advantage of, or exploits market fluctuations in order to make a profit. As a specialist in real estate, the property broker buys in order to resell. He must not be confused with the real-estate agent, who only acts as a middle-man between the different parties involved in a transaction, or with the developer, who has buildings constructed before selling them. The property broker may resell property bought as such or improved upon.

With deregulation, uncertainty and risks have reappeared in the real-estate markets, so market conditions conducive to speculation can develop more freely as the agents no longer have any point of reference. If suppliers overestimate the demand level, real-estate prices can rise to the point where they bear no relation to the price that solvent households or companies can pay. Then a speculative bubble appears and lasts so long as speculation increases and speculating agents exchange property between themselves, thus forgetting the investment and consumption dimension of real-estate property. Because of the price increases, householders seeking simply to satisfy their needs are excluded from the market.

CHAPTER 10

Case studies

10.1 Meylan ZIRST (Grenoble, Rhône–Alpes)

Introduction

The Meylan ZIRST (*Zone pour l'Innovation et les Réalisations Scientifiques et Techniques*) is an example of a technological park created at the beginning of the 1970s. It is a successful internationally renowned science park that covers 110 ha, houses 200 enterprises and has created 5,000 jobs. It was developed between the late 1960s and the beginning of the 1990s. Its interest lies in the decentralized process of development that took place during the 1970s: the idea was local, as was the development itself, and it succeeded because of the rationale of the process set in motion and the local forces that converged.

Central government was reluctant to aid the project, mainly for political reasons (the city of Grenoble was administered by the opposition parties at the beginning of the 1970s) and for policy reasons, since the priority of the *aménagement du territoire* policy was southeast France, the Sophia Antipolis zone. This local development shows how local forces can succeed when they have an entrepreneurial perspective.

The local and regional context

The Grenoble agglomeration, with 400,000 inhabitants in 1990, is ranked tenth in France, and is a medium-size agglomeration at the European level. Its industrial tradition has given it good economic performance in the past two decades, although the 1980s saw an economic crisis, but with strong recovery at the end of the decade. In the past two decades, the agglomeration has developed a specific link between industry and the universities based on diverse research activities, many of which have an international reputation (DATAR-RECLUS 1989, DATAR 1990b).

The rôle of local and regional authorities

The Grenoble agglomeration is located in a powerful French region,

Rhône–Alpes, and a prosperous department, Isère. The Rhône–Alpes region, second in size after the Paris region, used the economic crisis at the beginning of the 1980s to adapt the composition of its industries and services. Economic growth in the region was higher than the average French growth rate for 1986–9 (*Le Monde* 1991). The Rhône–Alpes region also has a positive demographic balance: employment and population in the Isère department have increased during the 1980s.

Local authorities play an important economic rôle in the Rhône–Alpes region and the Isère department. In Isère in particular, local authorities for at least a decade provided an assistance system for companies that focuses on land and buildings. The objective is to support employment. This assistance can be indirect (loan guarantees, discount on rents and market value of the industrial properties) or direct (loans, advances, interest allowances).

Local authority actions in Isère mostly involve land purchases and discounts on the market value of the industrial properties (62 per cent of total aid). Municipal assistance is preponderant (40 per cent of total assistance), particularly in land purchases (93 per cent). The local actions are sometimes on the edge of legality: the discounts for example on rents or market values of properties are accepted only to a 25 per cent limit, inside the special development areas (*aménagement du territoire* classified areas). In Isère, FF4,475,902 was spent in indirect assistance in 1985, FF18,437,336 in 1986 and FF27,328,600 in 1987.

Land and property markets of the Grenoble agglomeration

It is quite easy to describe the property market in the Grenoble agglomeration at the end of the 1980s, but it is more difficult to obtain appropriate information on the 1962–85 period, so this presentation is limited to the recent period. The supply of industrial buildings in 1987 was quite steady in comparison with polyvalent buildings, whose supply was booming. The polyvalent (multipurpose) buildings market more than doubled in size in 1987 and represented 90 per cent of total projects. The share of new buildings has risen (+16 per cent annual growth), and has had a strong influence on the growth of floor-space and the legal use of buildings through diversification. The average building on the market is an old building available to rent, with a floor area of between 500 m² and 1,000 m².

Renting developed strongly at the end of 1980s (see Table 10.1), and sales of industrial buildings also rose strongly in the mid-1980s from 6 per cent in 1985 to 40 per cent in 1986, mainly for important developments of over 2,000 m².

In 1989 the Grenoble chamber of commerce considered that the city's industrial property market was relatively short of polyvalent buildings, even

Table 10.1 Rental transactions in Grenoble (%).

	Industrial buildings	Polyvalent buildings
1986	60	82
1989	78	84

though the total stock of industrial buildings rose with industrial changes. The letting of industrial buildings decreased in 1989 by comparison with 1988: 38 industrial building units of $500\,m^2$ were let in 1989 (including 10 units of $600\,m^2$ in the town centre), compared with 66 units of $224\,m^2$ in 1988.

The rise in demand has not been satisfied because of the lack of land reserves on the outskirts of the agglomeration. Land prices are also too high for many companies, and existing buildings are usually too old and of inappropriate design. Municipalities also have a tendency to reject heavy industry in order to attract light activities that specialize in high technology and have good financial guarantees. Rents in the Grenoble agglomeration are a function of the category, age, areas and size of buildings the companies are looking for. Prices in 1989 are shown in Table 10.2.

Table 10.2 Rents for buildings in Grenoble.

	FF per m^2
Offices	450–500
Industrial buildings	280–300
Warehouses	230–280

Source: BIEN L'immobilier d'entreprise dans l'agglomeration Grenobloise [consultant's report] (1989).

The industrial property stock in the Grenoble agglomeration is becoming older, and many of the buildings will have to be completely rebuilt. There is a strong demand for new medium-size buildings ($80–100\,m^2$).

Planning instruments in Grenoble and the Meylan science park
Planning instruments in the Grenoble agglomeration The planning instruments used in the Grenoble agglomeration fall into two categories: planning instruments (*Schéma Directeur* – SD; and *Plan d'Occupation des Sols* – POS) and operational tools (*Zone d'Aménagement Différée* – ZAD; *Zone d'Aménagement Concerté* – ZAC; and *Société d'Economie Mixte* – SEM). All these instruments have been extensively used in Grenoble, and particularly in the case of the ZIRST project.

The master plan (SD) was prepared in 1969–73 by 23 communes, and was under review during 1990–91. Population was expected to increase to

700,000, with the creation of 1,700 ha of new activity zones (2,640 ha overall). The local plans (POS) were prepared roughly in accordance with the SD: the total area was much the same, with a few adaptations (Freschi & Freschi 1986). The land reservations needed by those developments were obtained through three deferred development areas (ZAD), particularly the ZAD of Meylan Montbonnot.

The surface areas allocated for development have not yet been fully developed. Out of a total of 1,700 ha, 602 ha were developed in 1984, with 350 ha in zones of activity. Most of the area (82 per cent of the surface area) was developed using the ZAD planning instrument, and the dominant developer was the *Société d'Aménagement et de Développement de l'Isère* (SADI) which dealt with 73 per cent of the surfaces developed, a typical French mixed (public–private) company. The ZIRST case is, in this respect, a good example of the use of those planning instruments.

Planning instruments in the Meylan Science Park   The idea for the ZIRST project was first formulated into a concrete project at the very beginning of the 1970s as a means of promoting high technology innovative products and providing an interface between industry, research and the university. The choice of the Meylan commune was the result of the priority given to environmental quality. It is in a residential suburb located close to the university. It was initially a private initiative, but was developed through strong local authorities policies without aid from central government.

To obtain space in ZIRST, an application must be made to a special management committee responsible for implementing and achieving the objectives of the development. Most companies on the site were created in the park itself (60 per cent) or are divisions or subsidiaries of large companies (39 per cent). Industrial users (apart from services) include artificial intelligence (3 per cent), robotics (5 per cent), remote and off-line process control (6 per cent), automation (7 per cent), training (11 per cent), software (13 per cent), industrial electronics (17 per cent), sensors (components and complete equipment) (17 per cent), and computers and peripheral activities (21 per cent).

Analysis of the development process 1969–91

The idea of the project emerged during the buoyant period of the late 1960s in Grenoble: a period of experimentation with new urban ideas and of the 1968 Olympic Games. The first full expression of the project is in the *Livre Blanc* of the *Schéma Directeur d'Aménagement et d'Urbanisme* in 1969, elaborated by the *Agence d'Urbanisme de l'Agglomeration Grenobloise*. It is the result of a strong process of discussion and concerted effort in the city of Grenoble, which gradually came to include local civil servants. The main

144

idea was to develop a link between the university and industry through research. The site chosen was near the university in a green suburban location (Bernardy de Sigoyer & Boisgontier 1988).

In 1971, the *Association pour l'Etude de la* ZIRST (AZIRST) was created and this elaborated the activity zone project by creating two institutional structures to implement it: the *Union pour la Promotion de l'Innovation dans la Region Grenobloise* (UPIRG) and the *Association pour la Promotion de la* ZIRST (PROZIRST). After the development of the idea at the end of the 1960s, a ZAD of 200 ha was created in 1971 to prevent any speculative land-banking.

The area was developed from 1971 under the ZAC procedure, with a concession to SADI. The area was classified as an activity zone in the *Schéma Directeur d'Aménagement et d'Urbanisme* in 1973, and defined in the POS of Meylan as an NA zone (100 ha, *coefficient d'occupation des sols* = 0).

SADI became the *aménageur*, with reversion of the buildings and facilities to the commune at the end of a 15-year period. SADI bought the land and serviced it. The commune gave it a financial guarantee and also had an influence on the delivery of construction permits. The particularity of the ZIRST lies in the fact that the municipality has given the choice of admitting enterprises to an agreement committee.

The promotion of the zone was the responsibility of a specialized company, PROZIRST, helped by public and private actors (chamber of commerce and banks). PROZIRST is a limited company created in 1972 to promote the zone, and also to pool resources in the site and to find solutions to problems that confront all companies. New approaches to developing new industries came in the 1970s with the creation of lettable premises on leases for terms of less than two years. The LOCAZIRST properties belong to the commune of Meylan, with a management mandate given in 1986 to a specialized company, the France-Régie company. This company is in charge of letting and managing four zones, each of 3,000 m².

One of the most important points in the development of this activity zone has been the creation of the agreement committee, in charge of selecting enterprises admitted in the area. This strongly adheres to its guidelines, which were designed to give the enterprises a common entrepreneurial spirit directly linked with innovation in new technologies.

The development on Montbonnot commune

ZIRST was developed at the end of the 1980s in the Montbonnot Commune as planned in the 1970s, on 45 ha. The plan provides for floor-space of 300,000 m² to be developed from 1990 to 2000. Developed land prices are high (FF220 per m2 in 1991). Servicing of the land began in 1989, and in March 1991 three firms arrived.

Table 10.3 The development process of the Meylan Montbonnot high technology park.

1969–73:	SDAU elaboration
1969:	Livre Blanc the idea of the ZIRST
1970:	ZAD Meylan Montbonnot created
1971:	SADI gets the concession for the project; AZIRST is created to define a green high technology park
1972:	PROZIRST created to promote the park; UPIRG (agreement committee) creation
1973:	ZAC Meylan creation (40 ha, first stage)
1977:	Inter Enterprise restaurant is open
1983:	Montbonnot POS adopted; Meylan POS adopted
1985:	ZAD Meylan Montbonnot reconducted; Montbonnot is start of the development process
1988:	Public enquiry and DUP on Montbonnot site
1989:	ZIRST Meylan complete; land acquisition on the Montbonnot site
1989–90:	Montbonnot commercialization
1990:	Development of 12 ha of the Montbonnot site

SADI (replaced in 1989 by *Grenoble Isère Développement*) has the concession for the aménagement. It has been supported by a private company, the Michel Ferrier group, which is responsible for development. The construction of buildings was given to private promoters. For each property development project, a property company was created with the Michel Ferrier group and banks. This holding company is also in charge of the creation of a business and service centre. The letting was given to national consultancy companies: Thouard, Bourdais and SODEC (a subsidiary company of the Caisse des Dépôts et Consignations).

The influence and efficiency of development and planning instruments
The planning instruments (SDAU, POS) are flexible and dependent in fact on the will of local authorities, particularly with respect to the development processes. The buoyant context of the late 1960s and early 1970s facilitated in this respect the elaboration of an agglomeration strategy which is less evident at the beginning of the 1990s. The activity zone has been very successful not only because of the number of enterprises and jobs created but also in reaching three main objectives: to develop a high technology park with innovative enterprises, to maintain an excellent environmental quality, and to submit developments to severe quality criteria or constraints.

One of the main problems the commune had to face at the beginning of the development process was the attitude of SADI, which perceived Meylan ZAC as developing more as an ordinary industrial zone than as a "Green high technology park". The Meylan commune had to exert considerable pressure to maintain its objectives. The second problem was that the enterprises themselves were not satisfied with the environmental and architectural constraints. A good deal of work was needed to explain their necessity if the park was to succeed (when explanations were not enough, the rules in the PAZ were there to be applied).

From the environmental point of view, the PAZ has been a most powerful instrument: it has incorporated all the environmental protection rules concerning, for example, trees, hedgerows and ditches. It has been fundamental also in giving a green quality to the site. Moreover, there was a general emphasis in the attitude of developers in favour of including environmental considerations in the development strategy. The buildings have also been carefully designed, with architectural rules applied by the Meylan commune.

The PAZ was organized flexibly to allow a process of adaptation during the 15 years of development, within the general characteristics of a Green high technology park. The image of ZIRST is very good at national level and very attractive for many companies. The image of the agglomeration has also benefited this development process. The impact of the high technology park on the Meylan commune is fundamental. At the end of the 1960s this commune was faced with the prospect of becoming a purely residential suburb. The large number of enterprises and jobs created has transformed it into one of the secondary centres of the Grenoble agglomeration, with a good balance of jobs and housing. The impact on the municipal budget is very good: Meylan is one of the richest communes of the department. Property and land prices have also increased quickly in that area, and are the highest of the agglomeration at between FF500 and FF700 per m^2 in 1991.

10.2 ZAC Citroën-Cevennes

This case study deals with the birth and the evolution of a large development scheme. This comprehensive development area (ZAC) is located in the southwest of Paris in the 15th district. The CDA is called Citroën-Cevennes after the former car factory, which relocated in the 1970s to the west of France.

The framework
The population of the Ile-de-France region is approximately 10,660,554 (1990 census). This is an increase of 580,000 people or 5.75 per cent since 1982 (see Ch. 1.3). Some districts grew: for example, the 18th, 10th, 11th and 20th districts. The density is 20,444.75 inhabitants per km^2. The Paris metropolitan area includes about 20 per cent of the national population on 2 per cent of the territory.

All of these demographic factors, plus the changing economic base because of the decline in manufacturing, affected Paris. Since 1970, the city council of Paris and the government has tried to solve them by improving housing. A new phenomenon began to throw this regional organization off balance: an east–west disequilibrium, with most office space being built west of Paris whereas housing was built to the east. Two out of the three major French car-makers had important factories in Paris (Citroën) or in the near suburbs (Renault in Boulogne-Billancourt). In the early 1970s both firms decided to relocate these factories.

The seeds of the large development scheme called *Zone d'Aménagement Concerté Citroën-Cevennes* lay in these events. It should be emphasized that the neighbouring area, between the Eiffel Tower and the Citroën area, had just undergone major redevelopment, including 13 high-rise buildings (*Front de Seine Développement*), and that this scheme was being criticized for various reasons.

The city council's instruments

An area of more that 21 ha available in a good location inside Paris is very unusual, and Paris city council has taken the initiative of using various legal instruments to control this development:

○ The land company is a kind of private property company financed by the city council, which works solely for the council. It does all the studies, prepares the policy and manages a part of its implementation. It has a great deal of responsibility, but acts only in the framework of city council's decisions, and it is totally subordinated to it.

○ The trilateral committee of co-ordination had a composition defined by the prefect of Paris. It included members of the Companie foncière, members of the city council and elected people from the district. This committee had the responsibility of providing the coordination between the different boards involved in the process.

○ The *Agence Foncière et Technique de la Région Parisienne* (AFTRP – Paris region land agency) is a regional public land bank that helped towns control urban development through technical assistance and financial subsidization for land purchase (see Part II).

○ The SEMEA XV is a typical French public–private company (*Société d'Economie Mixte* – SEM). It was created by the city council of Paris. SEM and city councils are frequently managed by the same people. SEMEA XV is the result of the transformation of a land company. It can use, through delegation of authority, the major tools of the city council, including expropriation and pre-emption.

In the case of the comprehensive development area of Citroën-Cevennes ZAC, this delegation was total, and its accounts are controlled by legal organizations. In this case, delegation included responsibility for preparatory studies, land purchase, compensation and rehousing, land preparation and development, technical control, marketing and sales to private companies.

Description of the site
The 15th district is quite near the centre of Paris and is very well linked by public and private transport systems. The city's most important landmark, the Eiffel Tower, overlooks the district. The Seine forms the northern border, with the Javel harbour belonging to the *Port Autonome de Paris*). Nearby, there is the *Beau Grenelle* centre with hotels and luxury shops, the Paris Exhibition Centre (*Porte de Versailles*) on a 36 ha site, and a sports centre. The district is very well linked with the other part of Paris by several important roads, four metro stations, three bus stations, and two RER (suburban rail) stations.

The project
The Citroën-Cevennes operation began in 1970–1 with impact and feasibility studies. The chronology and organization is set out in Table 10.4.

Table 10.4 Timetable of the Citroën-Cevennes operation.

1972:	A ZAD is created to limit the growth of land prices in order to control speculation; a ZAC is proposed to the prefect; the principles of the programme are studied; a wide-ranging agreement is prepared between the city council and AFTRP, with a large scope.
1976:	Preparation of a *plan de masse* (local scale plan), and a (PAE) a special exaction area (see Ch. 3.1).
1977:	An agreement (which followed the previous one between the Paris city council and AFTRP) is signed between the city council and SEMEA XV. This SEMEA XV is in charge of the comprehensive development area, and of managing the studies.
1979:	A forecast financial statement is made.
1980:	A report about the development is published.
1981:	The prefect allows the SEMEA XV to use expropriation, so they can easily buy land.
1982:	The SEMEA XV is chosen to manage the overall realization of the project (in France, the company undertaking preparatory studies is not normally responsible for implementation).

The land
In 1972, an agreement was signed between President Pompidou and the Citroën company. The factories were to move out of Paris but the company wanted some compensation and authorization to set up its headquarters in this part of the 15th district. The land price was estimated at FF500 million for 21 ha and the city council decided to grant a FAR (floor area ratio) of 3.

149

It valued the land at FF3,000 per m². Citroën agreed to sell its land for only FF1,759 per m² in return for some considerations. The city council decided to finance this through a subsidized loan of FF275 million from the *Caisse des Dépôts et Consignations* and FF100 million to be borrowed on the financial market at 9–10 per cent over 20 years.

The government decided to offer the city of Paris financial support by giving special subsidies of 40 per cent of the price of the land for green areas and 30 per cent of the cost of public equipment. The total is FF123 million. The government allowed the Citroën company to build its new headquarters on a 15,000 m² site, thereby enabling the company to build a floor area of 35,000–45,000m². In 1975, 45 per cent of the land was bought; in 1976 this was built up to 87 per cent, in 1977 94 per cent, and in 1978 100 per cent.

Acquisition was carried out by AFTRP with the agreement of the prefect. Its cost included financial charges of over FF45 million. But this financial cost and the management of the operation was seriously disputed by the opposition in the city council. Some interesting criticisms were made, to the effect that in this operation Citroën made a profit of FF300 million, having bought the land in 1910 for only FF1.5 million at 1972 prices; that only 1,000 social housing units were planned, although there were more than 1,200 households needing rehousing and 4,000 dilapidated dwellings in the district; and that Citroën was renting 14,434m² including 3,636m² that belonged to the city of Paris.

Table 10.5 Origins of land in Citroën-Cevennes.

	ha
Citroën factories	21.3
Municipal coach transports	1.7
Technical department of the town	1.4
French national railway company	8.0
Miscellaneous	2.6
TOTAL	35.0

Source: Municipality records.

The decision to leave the 15th district came from Citroën's management. In spite of thus being responsible for the breach of the letting lease, Citroën has been compensated. The city council replaced factories with offices, which was not good for the local employment market. Many people observed that there was a contradiction between the stated intention of decentralizing tertiary activities and the real policy and actions taken by Paris city council. The Citroën-Cevennes ZAC contributed to such contradictions with the modifications of the master plan.

Analysis of the project

The three key objectives of the project were to reverse the depopulation phenomenon, to provide some parks and other green spaces and public facilities in a district where these were lacking, and to develop the banks of the Seine. The latter was most important because for many years the policy of the city council had been to improve the Seine's banks, revitalizing them by creating public pathways. The river is a very important symbol both historically and geographically for Parisians, so the logic was to enable the inhabitants to see it and to live with it.

An important step in the process was the approval of the implementation scheme or PAZ (see Ch. 3.1). The PAZ is the synthesis of all regulations, as all the details of the programme are described in it. For the Citroën-Cevennes ZAC it included a green space of 13 ha looking onto the Seine, a large housing programme with 60 per cent social housing, several public facilities and some small factories and craft shops, with no more than 20 per cent for offices and tertiary activities.

The *Plafond Légal de Densité* (PLD – legal density ceiling, see Ch. 3.1) was fixed initially (after 1975) at 1.5 and exerted an influence since the FAR (COS or *Coefficient d'Occupation des Sols*) was higher in most parts: 2.7 for housing, 1.3 for tertiary activities, 3.0 for public facilities and retail. The development rights were sold to various developers at market prices, with some subsidies for social housing.

Implementation

There are four parts to the project: housing, public facilities, leisure activities and offices. The housing programme was planned to include 2,400 flats, with a large proportion of social housing. The development rights have been sold for FF12,000 per m^2 of floor area.

The programme of public facilities included a new hospital of 600–700 beds, a telecommunications centre of 10,000 m^2, and 14,800 m^2 for sports and public facilities, schools, etc. These buildings are rented to different

Table 10.6 Housing built at Citroën-Cevennes.

Flat size (rooms)	Number	m^2	%
1	101	18	7
2	296	46	21
3	486	60	35
4	356	73	26
5+	148	88	11

Source: Municipality records.

people or companies at low prices which are fixed by the city council at not more than FF600 per m².

The PAZ includes 67,000 m² of office space, of which 15,000 m² is for Citroën and 5,000–6,000 m² for the harbour administration offices. An important financial group wants to build 55,000 m² of office space, and some other buildings have been planned such as hotels and different buildings for small companies.

Implementation of the project was planned in five stages. The first stage included 5 ha for green spaces, 500 flats, a new hospital and office space. The second comprised the beginning of the park with underground parking for 1,000 cars and activity space, plus 700 flats. The third and fourth stages were completion of the park and 1,400 flats. The fifth stage was the end of the project, with 400 flats and a garden.

Financial aspects

Some indications of prices and costs can be given to illustrate the financial basis of the project. The land was bought at approximately FF3,000 per m² at a valuation for offices and small enterprises, and was sold at FF11,000 per m². Thus the city council made a profit of FF7,000 per m². On completion, the users bought the floor area at FF30,000–35,000 per m². Meunier Promotion, a private property developer, bought land on behalf of a financial and banking group at FF5,000 per m², 55,000 m² of which was bought in one block. It probably sold its investment at the end of the programme at FF40,000 per m².

Social housing programmes had to be heavily subsidized to cover the very high price of development rights. PLA (subsidized loans for rental housing) carry an interest rate of 6.89 per cent over 34 years. PC loans cover 90 per cent of the reference price with a variable interest rate. PLI, subsidized loans for middle-income rental housing, represent 70 per cent of the reference price with an interest rate of 8.4 per cent over 25 years. PLA loans financed 45 per cent of the housing programme, 5.5 per cent by PC loans, 26.0 per cent by PLI loans and 23.5 per cent of the housing programme was undertaken without any subsidy. It should be emphasized that the difference between the rental levels for social housing and private housing fluctuate by factors of up to six inside the Citroën-Cevennes ZAC, and that frequent criticism is levelled at the city council concerning the allocation of social housing.

Conclusion

In this operation, the government played an important part. Yet the institutional maze that characterizes the city of Paris makes it difficult to assess the level of state control over Paris city council. The beginning of the Citroën ZAC took place in a favourable political context since Paris was governed at that time by the party in power nationally. Mitterrand's election in 1981 modified the situation, and since 1982–3 decentralization has changed the relationship between the government and Paris, which is always distinctive because Paris is the capital. Delay in the creation of the Citroën ZAC came from a double reform in 1977: the modification of the Paris statute and the transformation of the ZAC implementation process. The operation should have been finished by now, but in fact at least two years are still required for the completion of all buildings and public facilities. About 20 years will therefore have elapsed between the decision to relocate the Citroën factory and the complete renewal of the area, so implementation of this ZAC has been a long process. Some additional points to note concerning this ZAC are that the city council controls the process and programming through the private public company SEM XV, thus receiving part of the added value resulting from the process of transformation of the site, and that it has not resolved the question of depopulation and social mixture since a gentrification process seems to have taken place.

PART IV
Evaluation

CHAPTER 11

Evaluation of
urban land and property markets

11.1 Evaluation of the functioning
of the urban land market

To assess the functioning of this land market, with regard to certain aims and criteria, it is important to approach the issue from different perspectives. From the political point of view, the decentralization period has brought about major changes in the distribution of powers at different levels. The responsibility for preparing, publishing, approving and implementing local plans has produced growing planning powers for the municipalities. This probably improves the efficiency and the adaptation of the local planning system to uncertainty and economic changes. Public land-banking has not been a conclusive success in France, except in two cases: new towns and particular cases such as in Rennes.

However, innovative tools such as the pre-emption right allow the implementation of comprehensive planning in specific contexts (ZAC, ZAD procedures, etc.) and it has activated land planning. In addition, expropriation is mostly used for projects of public interest or national defence. This process involves complex and precise legal procedure, including a public inquiry. Consequently, flexible zoning has resulted in some variable outcomes. (cf. II.3 for a more detailed analysis) On one hand, there are elaborate projects (cf. ZAC Citroën case study). On the other hand, abuse by local authorities cannot be challenged effectively by administrative courts (cf. II.1.1). As deregulation overruns legislation, current urban planning laws are the result of a succession of statutes and decrees that are unable to achieve coherent land development.

Moreover, permits to build or to subdivide, among others, are now granted by local authorities if a local plan has been put into effect. If this is not the case, the state retains this competence. The main difficulty lies in the fact that various authorities lack certain knowledge of the economic consequences of planning on land markets. Therefore, land prices adjust to this classification since developer/builders seek land in other less expensive areas; this phenomenon can be labelled leap-frogging. Since there is no compensating fiscal treatment to land, land prices depend strongly on the local plan's prescriptions or eventual changes. This factor explains the commonly held perception of unfair planning at the local level. Likewise, public participation is often minimal. For example, some relevant information is often withheld when doing environmental impact studies or public inquiries.

Another failure stems from the growing inability of the land law system to cope with the different consequences of urban development. The complexities and inconsistencies in this system require drastic simplification since a single plot of land is submitted to several different regulations. In addition to this fact, land taxation, which is one of the four basic taxes at the local level, is considered to be inadequate and inefficient to undo land-hoarding. While there appears to be an inherent paradox, the decentralization move versus the tax system's immobility, this transition period has brought into question the whole decision-making process.

To sum up, the issue in France is no longer whether the public ownership of urban land or the free-market framework is better for urban development, but how to carry out a sound land policy that will take into account the linkages between planning and market mechanisms.

11.2 Evaluation of the functioning of the market for urban property

Nowadays, the urban property market is almost deregulated, yet some questions still remain. Should the side effects of deregulation be mitigated, especially the socially exclusionary processes that result from the free-market mechanism? At the individual level, it appears as if the needs of those commanding ineffective demand (the disfavoured, low-income households, etc.) are being increasingly excluded from the urban property market.

With deregulation, economic agents are not guided by controls designed to ensure that their development meets public expectations, so speculation can develop in areas of high urban pressure. In spite of this background, we cannot conclude by saying that the functioning of the urban property market is bad. Deregulation has been one of the government's prerogatives during the past 10 years and it is necessary to take into account the consequences of this policy. On the macroeconomic side, the property boom, from 1987 onwards, can be further evidenced by the impressive increase of loans granted to real-estate brokers and developer/builders (See Table 11.1).

In 1991, the stabilization or even the first evidence of incipient decline of

REFERENCES

ADEF 1983. *Les enjeux de la fiscalité foncière*. Paris: Economica.

ADEF 1989. *Outils fonciers, modes d'emploi*. Paris: ADEF Editions.

ADEF 1990. *Un droit inviolable et sacré: la propriété*. Paris: ADEF Editions.

ADEF 1991. *Security, fluidity and transparency of land and property transfers*. Proceedings of the ADEF Conference. Paris: ADEF.

Aicardi, M. 1988. *La fiscalité du patrimoine. Rapport présenté au Ministre de Finances*. Paris: La Documentation Française.

Alterman, R. (ed.) 1988. *Private supply of public services: evaluation of real estate exac tions, linkage and alternative land policies*. New York: New York University Press.

Ardant, G. 1971. *Histoire de l'impôt*, 2 vols. Paris: Fayard.

Auby, J. B. & J. F. Auby 1990. *Droit des collectivités locales*. Paris: Presses Universitaires de France.

Auguste-Thouard 1990. *Conditions de développement du terrain de Pompey*. Consultants' report. Paris: Auguste-Thouard.

Babeau, A. 1988. *Le patrimoine aujourd'hui*. Paris: Nathan.

Bernardy de Sigoyer, M. & P. Boisgontier 1988. *Grains de technopoles: micro-entreprises grenobloises et nouveaux éspaces productifs*. Grenoble: Presse Universitaire Grenoble.

Besson-Guillaumot, M. 1986. *Town and country planning in France*. In *Planning law in Western Europe*, 2nd edn, J. F. Garner & N. P. Gravells (eds), 152–87. Amsterdam: Elsevier.

Booth, P. 1991. The theory and practice of French development control. In *Town planning responses to city change*, V. Nadin & J. Doak (eds), 161–73. Aldershot, England: Avebury.

Carpentier, C. 1988. La ZAC Citroën-Cévennes: un example de la politique d'aménagement d'urbanisme menée par la ville de Paris. Master of Planning dissertation, Université de Paris IV (Sorbonne).

Chaline, C. 1987. *Major land-owners in France and Great Britain*. Paris: ADEF Editions.

Chapuy P. M. B. 1984. France. In *Planning in Europe: urban and regional planning in the EEC*, R. H. Williams (ed.), 37–48. London: Allen & Unwin.

Comby, J. & V. Renard 1985. *L'impôt foncier*. Paris: Presses Universitaires de France.

Comby J. & V. Renard (eds) 1990. *Land policy in France*. Paris: ADEF.

CREDOC 1988. *Opinions et aspirations en France: 10 années d'observation*. Centre de Recherche pour l'Etude et l'Observation des Opinions de Vie (CREDOC), 43. Paris: CREDOC.

CREDOC 1989. *Conditions de vie et aspirations de Français*. Centre de recherche pour l'étude et l'observation des opinions de vie (CREDOC), Rapport 65. Paris: CREDOC.

DATAR 1990a. *Rapport interministériel sur la politique foncière dans les anciens tissus industriels*. Paris: DIV-DATAR.

DATAR 1990b. *Vingt technopoles: un premier bilan*. Paris: La Documentation Française.

REFERENCES

DATAR-RECLUS 1989. *Les villes européennes*. Paris: La Documentation Française.

Denis, M. 1973. *Rennes au XIX siècle: ville parasitaire*. Rennes: Annales de Bretagne No. 2.

Direction-Générale des Collectivités Locales 1989. *Inventaire général des impôts locaux*, 2nd edn. Paris: La Documentation Française.

Direction-Générale des Collectivités Locales 1991. *Guide statistique de la fiscalité direct locale*. Paris: La Documentation Française.

Douenel, C. 1990. L'aménagement des berges de la ZAC Citroën-Cévennes, symbole de la réconciliation de la ville avec son fleuve. Master of Planning dissertation, Université de Paris IV (Sorbonne).

Freschi, L. & N. Freschi 1986. Les zones d'activité de la région grenobloise. *Revue de Géographie Alpine* **74**, 247–64.

Granelle, J-J. 1970. *Espace urbain et prix du sol*. Paris: Sirey.

Granelle, J-J. & S. Guelton 1991. Récentrage des marchés. Etudes Foncières **50**, Mars.

Gravier, F. 1947. *Paris et le désert français*. Paris: Flammarion.

Guengant, A. 1989. *Les nouveaux coûts d'urbanisation*, 3 tômes. Paris: Plan Urbain et Ministère d'Affaires Urbain.

Guigo, J. L. & J. M. Legrand 1986. *Fiscalité foncière: analyse comparée des pays de l'OCDE*. Paris: Economica.

INED 1991. *Population et sociétés*. Bulletin of the Institut National des Etudes Demographiques. Paris: INED.

INSEE 1986. *Economie Lorraine*. Rapport No. 45, Institut National des Statistiques et de l'Etudes Economiques. Paris: INSEE Publications.

INSEE 1989. *Comptes et indicateurs économiques: rapport sur les comptes de la nation 1989*. Institut National des Statistiques et de l'Etudes Economiques. Paris: INSEE Publications.

INSEE 1991. *Données sociales 1991*. Institut National de la Statistique et l'Etudes Economiques. Paris: INSEE Publications.

Klein, J. S. 1986. *L'explosion des impôts locaux*. Paris: La Documentation Française.

Lacaze, J-P. 1989 *Les Français et leur logement: éléments de socio-économie de l'habitat*. Paris: Ecole Nationale des Ponts et Chausées Press.

Lefebvre, B., M. Mouillart, S. Occhipinti 1990. *Perspective du secteur et du financement du logement en France à l'horizon 1993*. Proceedings of Conference on Building and Planning, Assemblé Nationale, Paris.

Lefebvre, B., M. Mouillart, S. Occhipinti 1992. *La politique du logement en France: 50 ans pour un échec*. Paris: L'Harmattan.

L'Hardy, P. 1990. *Epargne des ménages, montée des placements*. INSEE Première No.105 Paris: INSEE.

Monzavi, M. 1988. ZAC Citroën-Cévennes. PhD dissertation, Université de Paris IV (Sorbonne).

Merlin, P. 1982. *Les transports à Paris et en Ile-de-France*. Paris: La Documentation Française.

Motte, A. 1992. France. In *Industrial property markets in western Europe*, B. Wood &

REFERENCES

R. H. Williams (eds), 97–127. London: E. & F. N. Spon.

Mouillart, M. & Y. Martin 1989. *Parc locatif privé et politique du logement, un essai de bilan prospectif.* Proceedings of Conference on Trois siècles de logement des français, Paris.

OECD 1990. *Urban land markets in the OECD countries: final report.* Paris: Organisation for Economic Cooperation and Development.

Pearsall, J. 1988. France. In *Land and housing policies in Europe and the USA*, G. Hallett (ed.), 76–98. London: Routledge.

Prest, A. R. 1981. *The taxation of urban land.* Manchester: Manchester University Press.

Punter, J. V. 1989. France. In *Planning control in western Europe*, H. W. E. Davies, D. Edwards, A. J. Hooper, J. V. Punter (eds), 149–51. London: HMSO.

Questiaux, N. 1989. *Les français et leur revenus: le tournant des années 80.* Centre d' études des revenus et des coûts (CREC). Paris: La Documentation Française.

Saglio, M. 1980. *L'offre foncière.* (Report to the prime minister on land supply). Paris: La Documentation Française.

Savitch, H. V. 1988. *Post-industrial cities: politics and planning in New York, Paris and London.* Princeton: Princeton University Press.

STU 1990. *Guide de l'aménagement.* Service Technique de l'urbanisme. Paris: Ministry of Urban Affairs Press.

Topalov, C. 1985.Prices, profits and rents in residential development: France 1960–1980. In *Land rent, housing and urban planning: a European perspective*, M. Ball, V. Bentivegna, M. Edwards, M. Folin (eds), 25–45. London: Croom Helm.

Tribillon, J-F. 1991. *L'urbanisme.* Paris: La Découverte.

Wilson, I. B. 1988. *French land-use planning in the Fifth Republic: real or imagined decentralisation?* Nijmeegse Planologische Cahiers 27. Nijmegen: Geografisch en Planologisch Instituut, Katholieke Universiteit Nijmegen.

SEE ALSO

Economie et Statistique. (quarterly review of the INSEE.) Paris: INSEE Publications, 209–401.

Etudes Foncières. (Quarterly review of the ADEF.) Paris: ADEF Publications, 31–52.

INDEX OF ENGLISH TERMS

INDEX OF FRENCH TERMS

Printed and bound by CPI Group (UK) Ltd, Croydon, CR0 4YY

22/10/2024

01777622-0020